Health and Safety at Work Revision Guide

For the NEBOSH National General Certificate

Ed Ferrett

Routledge
Taylor & Francis Group

LONDON AND NEW YORK

First edition published 2009 by Elsevier
This edition published 2012 by Routledge
2 Park Square, Milton Park, Abingdon, Oxon OX14 4RN

Simultaneously published in the USA and Canada
by Routledge
711 Third Avenue, New York, NY 10017 (8th Floor)

Routledge is an imprint of the Taylor & Francis Group, an informa business

Ferrett, Ed.
    Health and safety at work revision guide / Ed Ferrett.—2nd ed.
        p. ; cm.
    Includes bibliographical references and index.
    I. Title.
    [DNLM: 1. Occupational Health—Outlines.   2. Accidents, Occupational—prevention &
    control—Outlines.   3. Hazardous Substances—Outlines.   4. Safety Management—
    Outlines.   5. Workplace—legislation & jurisprudence—Outlines. WA 18.2]
    363.11—dc23

                                                                    2011050657

Pbk ISBN: 978-0-415-51979-3
Ebk ISBN: 978-0-203-11381-3

Typeset in Rotis Sans Serif
by RefineCatch Limited, Bungay, Suffolk

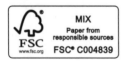

Printed and bound in Great Britain by
TJ International Ltd, Padstow, Cornwall

Contents

Preface

Welcome to the *NEBOSH National General Certificate Revision Guide*. The guide has been designed to be used together with the NEBOSH National General Certificate syllabus and the textbook *Introduction to Health and Safety at Work* by Hughes and Ferrett. The guide gives only a basic summary of the NEBOSH National General Certificate course, and a full explanation of all the topics is given in the textbook.

The Revision Guide has the following features:

■ It follows the latest NEBOSH National General Certificate syllabus;

■ Revision notes for each of the elements of the two units – NGC1 (Management of Health and Safety) and NGC2 (Controlling Workplace Hazards);

■ A summary of the learning outcomes and key points is given for each element;

■ Important diagrams are included to help revision;

■ A small section on 'Additional construction items', which is not in this syllabus but may be useful to Health and Safety managers;

■ There is a final section to advise on the preparation for the examination and some specimen answers to both long and short answer questions taken from recent NEBOSH examination papers.

The Revision Guide will also be useful to those who have specific health and safety responsibilities in their jobs and those who are studying other courses that include important health and safety elements, for example courses in Engineering, Business Studies, the Health Services and Retail and Hotel Management.

The compact size of the Revision Guide ensures that it can be easily carried and used for revision at any time or place. It should be used throughout the course alongside the textbook and course hand-outs.

Good luck with your studies.

List of Principal Abbreviations

ACM	Asbestos containing material
ACOP	Approved Code of Practice
AFFF	Aqueous film forming foam
CDM	Construction (Design and Management) Regulations
COSHH	Control of Substances Hazardous to Health Regulations
dB	Decibels
dB(A)	Decibel (A-weighted)
dB(C)	Decibel (C-weighted)
DSE	Display screen equipment
EU	European Union
HAV	Hand–arm vibration
HSE	Health and Safety Executive
HSW Act	Health and Safety at Work Act 1974
ILO	International Labour Organization
LEV	Local exhaust ventilation
MEL	Maximum exposure limit
MHSWR	Management of Health and Safety at Work Regulations
MIRA	Motor Industry Research Association
MSD	Musculoskeletal disorder

NEBOSH	National Examination Board in Occupational Safety and Health
OES	Occupational exposure standard
PPE	Personal protective equipment
PUWER	The Provision and Use of Work Equipment Regulations
RCD	Residual current device
RIDDOR	The Reporting of Injuries, Diseases and Dangerous Occurrences Regulations
RPE	Respiratory protective equipment
SPL	Sound pressure level
STEL	Short-term exposure limit
TWA	Time-weighted average
WBV	Whole body vibration
WEL	Workplace exposure limit
WRULD	Work-related upper limb disorder

Management of Health and Safety

Foundations in Health and Safety

1.1

LEARNING OUTCOMES

Outline the scope and nature of occupational health and safety ☐

Explain the moral, legal and financial reasons for
promoting good standards of health and safety in the workplace ☐

Explain the legal framework for the regulation of health
and safety, including sources and types of law ☐

Explain the scope, duties and offences of employers,
managers, employees and others under the Health and
Safety at Work etc. Act 1974 (HSW Act) ☐

Explain the scope, duties and offences of employers,
managers, employees and others under the Management
of Health and Safety at Work Regulations ☐

Outline the legal and organisational health and safety
roles and responsibilities of clients and their contractor ☐

KEY REVISION POINTS

The definitions of hazard, risk, civil law, criminal law, common law and statute law ☐

The employer's duty of care and other common law and statutory duties ☐

Criminal offences and defences ☐

Civil liabilities and defences (particularly negligence) ☐

The business case for health and safety (direct, indirect, insured and uninsured costs) ☐

The legal framework – the Health and Safety at Work Act, the Management of Health and Safety at Work Regulations, absolute and qualified duties ☐

The powers of the enforcement officer, health and safety offences, including corporate manslaughter and corporate homicide, and penalties ☐

The role and functions of other external agencies ☐

The health and safety responsibilities and duties of employers to their employees and others affected by their undertaking, such as contractors, general public, visitors, patients and students ☐

The health and safety responsibilities of directors, managers, supervisors, employees and the self-employed ☐

Duties and responsibilities of manufacturers and others in the supply chain ☐

The duties and responsibilities between client and contractor to ensure a high standard of health and safety during the contract ☐

The legal duties under the Construction (Design and Management) Regulations ☐

DEFINITIONS

- Accident – an unplanned event that results in damage, loss or harm
- Hazard – the potential of something to cause harm
- Risk – the likelihood of something to cause harm
- Civil Law – duties of individuals to each other
- Criminal Law – duties of individuals to the State
- Welfare – provision of facilities to maintain health and well-being of people in the workplace (e.g. washing, sanitary and first aid)
- Residual risk – remaining risks after controls applied
- Near miss – any incident that could have resulted in an accident
- Dangerous occurrence – a near miss that could have led to serious injury or loss of life
- Common Law – law based on court judgments
- Statute Law – law based on Acts of Parliament.

REASONS FOR GOOD HEALTH AND SAFETY MANAGEMENT

MORAL REASONS

Need to provide a reasonable standard of care and ethical reasons to reduce:

- accident rates
- industrial disease and ill-health rates.

LEGAL REASONS

Employers have a duty to take reasonable care of workers. Poor management can lead to:

- prosecutions
- civil actions – compensation claims.

FINANCIAL REASONS

Poor health and safety management can lead to:

- direct and
- indirect costs.

Good health and safety management can lead to:

- a more highly motivated workforce resulting in an improvement in the rate of production and product quality; and
- an improved image and reputation of the organisation with its various stakeholders.

SOCIAL REASONS

These include:

- societal expectation of good standards of health and safety; and
- duty of care (details under 'Aspects of civil law' below).

COSTS OF ACCIDENTS AND ILL HEALTH

DIRECT COSTS

Directly related to the accident and may be insured or uninsured.

Insured direct costs normally include:

- claims on employer's and public liability insurance;
- damage to buildings, equipment or vehicles; and
- any attributable production and/or general business loss.

Uninsured direct costs include:

- fines resulting from prosecution by the enforcement authority;
- sick pay;
- some damage to product, equipment, vehicles or process not directly attributable to the accident (e.g. caused by replacement staff);

- increases in insurance premiums resulting from the accident;

- any compensation not covered by the insurance policy due to an excess agreed between the employer and the insurance company; and

- legal representation following any compensation claim.

INDIRECT COSTS

Costs which may not be directly attributable to the accident but may result from a series of accidents.

Insured indirect costs include:

- a cumulative business loss;

- product or process liability claims; and

- recruitment of replacement staff.

Uninsured indirect costs include:

- loss of goodwill and a poor corporate image;

- accident investigation time and any subsequent remedial action required;

- production delays;

- extra overtime payments;

- lost time for other employees, such as a First Aider, who attend to the needs of the injured person;

- the recruitment and training of replacement staff;

- additional administration time incurred;

- first aid provision and training; and

- lower employee morale, possibly leading to reduced productivity.

Some of these items, such as business loss, may be uninsurable or too prohibitively expensive to insure. Therefore, insurance policies can never cover all of the costs of an accident or disease, because either some items are not covered by the policy or the insurance excess is greater than the particular item cost.

EMPLOYERS' LIABILITY COMPULSORY INSURANCE

- ▪ legal requirement for all employers;

- ▪ covers the employer's liability in the event of accidents and work-related ill health to employees and others who may be affected by their operations;

- ▪ ensures that any employee who successfully sues his/her employer following an accident is assured of receiving compensation irrespective of the financial position of the employer; and

- ▪ made available either by display or electronically at each place of business.

LEGAL FRAMEWORK FOR HEALTH AND SAFETY

Sub-divisions of Law	
Criminal Law	Civil Law
• enforced by the State to punish individuals (and/or organisations)	• disputes between individuals (and/or organisations) to address a civil wrong (tort)
• individual is prosecuted by an Agency of the State (e.g. Police, HSE, Local Authorities, Fire Authority)	• individual(s) and/or organisations are sued
• individual(s) guilty or not guilty	• individual(s) are liable or not liable
• courts can impose fine or imprisonment	• courts can award compensation and costs
• proof 'beyond reasonable doubt'	• proof based on 'balance of probabilities'
• cannot insure against punishment	• employers must insure against civil actions (**Employers' Liability Insurance**)

Sources of Law	
Common Law	**Statute Law**
• based on judgments made by judges in courts	• law laid down by Acts of Parliament
• generally courts bound by earlier judgments (precedents)	• Health and Safety at Work Act 1974
• lower courts follow judgments of higher courts	• specific duties mainly in Regulations or Statutory Instruments
• in health and safety definitions of negligence, duties of care and terms such as 'practicable' and 'reasonably practicable' are based on common law judgments	• takes precedence over Common Law

SUB-DIVISIONS AND SOURCES OF LAW

Figure 1 shows the relationship between the sub-divisions and sources of law

Sources of Law	**Sub-divisions of Law**
Common Law	Criminal Law
Statute Law	Civil Law

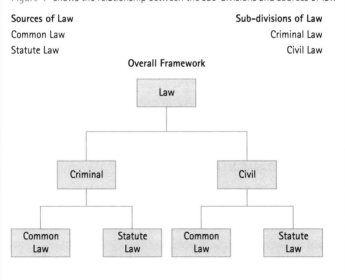

Overall Framework

Figure 2 shows the court system for health and safety in England and Wales (NB European courts advise the Supreme Court)

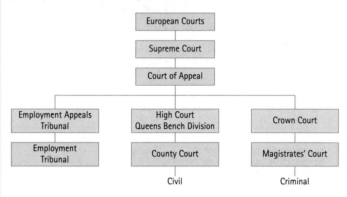

COMPARISON BETWEEN ENGLISH AND SCOTTISH COURTS

	Lower court	Higher court*
England and Wales	Magistrates	Crown
Scotland	Sheriff	High Court of Justiciary
Cases	Summary	Indictable

* Higher court deals with some summary cases referred to it by the lower court

ASPECTS OF CIVIL LAW

EMPLOYER'S COMMON LAW DUTY OF CARE

The employer has a duty of care to each of his employees. This duty cannot be assigned to another person. The duty of care falls into five categories and the employer must provide:

◼ a safe place of work, including access and egress;

◼ safe plant and equipment;

- a safe system of work;

- safe and competent fellow employees; and

- adequate levels of supervision, information, instruction and training.

Employees have a right not to be harmed in any way by their work and they are expected to take reasonable care in their workplace.

NEGLIGENCE

- lack of reasonable care or conduct resulting in injury, damage or loss

- it must be reasonably foreseeable that the acts or omissions could result in injury.

Defences against negligence claims	Partial defences against negligence claims
• A duty of care was not owed;	• Contributory negligence – employee contributed to the negligent act;
• There was no breach of the duty of care; and	• Volenti non fit injuria – the risk was willingly accepted by the employee;
• Any breach of a duty of care did not result in the specific injury, disease, damage and/or loss suffered.	• Acts of God, riot, terrorist event etc.

Negligence claims must be made within a set time.

VICARIOUS LIABILITY

When the defendant is acting in the normal course of his employment during the alleged incident, the defence of the action is transferred to his employer.

TORT OF BREACH OF STATUTORY DUTY

When there is a breach of statutory duty, such as the duties laid down by the HSW Act or its regulations, an individual who has been affected by the breach may pursue a civil action to obtain compensation.

Some parts of the HSW Act itself and its regulations specifically prohibit a civil action following a breach of that statutory duty – these breaches are termed **statute-barred**.

DOUBLE-BARRELLED ACTION

If an employee suffers an injury at his place of work, he may sue his employer both for negligence and the breach of a statutory duty. Such an action is termed a **double-barrelled action**.

LEVELS OF STATUTORY DUTY

Absolute	Practicable	Reasonably practicable
The employer MUST comply with the law. Regulations use the verbs 'must' or 'shall'	The employer must comply if it is technically possible. Difficulty, inconvenience or cost cannot be taken into account	If the risk is small compared with the cost, time and effort required to further reduce the risk, then no action is required

THE LEGAL FRAMEWORK FOR THE REGULATION OF HEALTH AND SAFETY

INFLUENCE OF THE EUROPEAN UNION (EU)

- Role is to harmonise health and safety standards across member states

- Powers of EU in health and safety mainly derived from the Treaty of Rome (1957) and the Single European Act (1986)

- Article 95A (was 100A) – health and safety standards for plant and equipment

- Article 138A (was 118A) – minimum health and safety standards in employment

- European Directives set out the specific minimum aims of a given area of health and safety

- European Directives must be incorporated into the national law of all member states

- First introduction into UK law was in 1992 with the 'European Six Pack'

- European Court of Justice rules on interpretation of EU law

- European Court of Human Rights (covers a wider area than just the EU) interprets the European Human Rights Convention.

HEALTH AND SAFETY AT WORK ACT 1974

BACKGROUND TOPICS

The Health and Safety at Work Act (HSW Act) was introduced in 1974 and is a criminal law based on the recommendations of the Robens Report.

MAIN RECOMMENDATIONS OF THE ROBENS REPORT

1 A single Act to cover all workers containing general duties
2 The Act should cover everyone affected by the employer's undertaking
3 Emphasis on management of health and safety, including training and supervision
4 Encouragement of employee involvement in accident prevention
5 Enforcement should be targeted at 'self-regulation' rather than prosecution.

The HSW Act is an **Enabling Act**, allowing regulations to be produced by the HSE on behalf of the Secretary of State without further Acts of Parliament being required.

The HSW Act contains mainly general duties, with specific duties defined in Regulations.

ROLE AND ACTIVITIES OF THE HSE

- advises on the development of regulations;

- enforces health and safety regulations;

- provides information to organisations (ACOPs, guidance notes, leaflets and other publications, accident and ill-health statistics);

- launches national health and safety campaigns on specific topics;

- undertakes accident and other investigations;

- offers advice to employers and others on statutory duties;

- takes enforcement action;

- instigates criminal proceedings and publicises organisations that receive enforcement notices.

REGULATIONS

- state the law

- often implement EU Directives

- breaches are criminal offences possibly leading to enforcement action

- describe the minimum health and safety standards that need to be achieved

- usually apply across all organisations but sometimes apply to specific industries.

APPROVED CODES OF PRACTICE (ACOPS)

- supplementary practical interpretation of regulations that gives more detail on the regulatory requirements

- special legal status – quasi-legal because it may be possible to comply with regulations by some other more effective practice

- ACOPs are legally binding if the particular regulations indicate that they are or if they are quoted in an Enforcement Notice.

GUIDANCE

- two forms of guidance – legal and best practice, both available as HSE publications

- not legally binding

- gives more informative and practical advice than ACOPs

- also often available as British Standards and as industrial or trade guidance.

THE HEALTH AND SAFETY AT WORK ACT

SECTION 2 DUTIES OF EMPLOYERS TO EMPLOYEES

General duty – ensure, as far as is reasonably practicable, the health, safety and welfare of all employees

Specific duties –

- safe plant and systems of work;

- safe use, handling, transport and storage of substances and articles;

- provision of information, instruction, training and supervision;

- safe place of work, access and egress;

- safe working environment with adequate welfare facilities;

- a written safety policy together with organisational and other arrangements (if five or more employees); and

- consultation with safety representatives and formation of safety committees where there are recognised trade unions.

SECTION 3 DUTIES OF EMPLOYERS TO OTHERS AFFECTED BY THEIR UNDERTAKING

- 'others' could include contractors, general public, visitors, patients, students.

SECTION 4 DUTIES OF LANDLORDS OR OWNERS

- safe access and egress for those using the premises; and

- buildings and any equipment supplied with them are safe.

SECTION 6 DUTIES OF SUPPLIERS

Suppliers (including designers) of articles and substances for use at work to ensure, as far as is reasonably practicable, that

- articles are designed and constructed to be safe and without risk to health at all times when they are being set, cleaned, used and maintained;
- substances are similarly safe and without risk to health when being used, handled, stored or transported;
- arrange, where necessary, for suitable testing and examination; and
- supply suitable safety information and any revisions to customers.

SECTION 7 DUTIES OF EMPLOYEES

- take care for the health and safety of themselves and others who may be affected by their acts or omissions; and
- co-operate with their employer, as far as is necessary, to ensure compliance with any statutory health and safety duty.

SECTION 8

No person to misuse anything provided for health, safety or welfare purposes.

SECTION 9

Employees cannot be charged for health and safety requirements.

ENFORCEMENT OF THE ACT

POWERS OF INSPECTORS

- enter premises at any reasonable time, accompanied by a police officer, if necessary;
- examine, investigate and require the premises to be left undisturbed;
- access to all records and other relevant documents;

- take samples, photographs and, if necessary, dismantle and remove equipment or substances;

- seize, destroy or render harmless any substance or article;

- take statements;

- issue improvement and prohibition notices and, possibly, a formal caution; and

- initiate prosecutions.

ENFORCEMENT NOTICES

Improvement notice

Issued for a specific breach of the law.

Appeal within 21 days to an Employment Tribunal – notice suspended until appeal is heard or withdrawn.

Prohibition notice

Issued to halt an activity that could lead to serious injury.

Appeal within 21 days to an Employment Tribunal – notice is not suspended.

Deferred prohibition notice – stops the work activity within a specified time limit.

Penalties	
Summary offences	**Indictable offences**
• for most health and safety offences up to £20 000 fine and/or up to 12 months' imprisonment	• unlimited fines for all health and safety offences
• up to 5 years' disqualification for convicted directors	• up to 2 years' imprisonment for all health and safety offences
	• up to 15 years' disqualification for convicted directors

Work-related deaths

- Investigated by police initially to ascertain whether a charge of manslaughter (culpable homicide in Scotland) or corporate manslaughter is appropriate.

- If there are no such charges, the HSE or Local Authority continues the investigation.

THE CORPORATE MANSLAUGHTER AND CORPORATE HOMICIDE ACT

An organisation will have committed the new offence if:

- it owes a duty of care to another person in defined circumstances;

- there is a management failure by its senior managers; and

- it is judged that its actions or inaction amount to a gross breach of that duty resulting in a person's death.

The health and safety duties relevant to the Act are:

- employer and occupier duties including the provision of safe systems of work and training on any equipment used; and

- duties connected with:

  - the supply of goods and services to customers

  - the operation of any activity on a commercial basis

  - any construction and maintenance work; and

  - the use or storage of plant, vehicles or any other item.

A breach of duty is gross if the organisation's management of health and safety falls far below that which would reasonably be expected.

On conviction, the offence is punishable by an unlimited fine and the courts will be able to make:

- a remedial order requiring the organisation to take steps to remedy the management failure concerned;

- a publicity order requiring the organisation to publicise details of its conviction and fine; and

■ an order to pay the costs of the prosecution and possibly compensation for bereavement and funeral expenses.

The publicity order requires an organisation convicted of corporate manslaughter to publicise:

■ the fact that it has been convicted;

■ the particulars of the offence;

■ the amount of any fine;

■ the terms of any remedial order.

ROLE AND FUNCTIONS OF EXTERNAL AGENCIES

FIRE AND RESCUE AUTHORITY

■ enforces fire safety law;

■ undertakes random fire inspections (often to examine fire risk assessments);

■ can issue alteration, improvement and prohibition notices; and

■ needs to be informed during the planning stage of building alterations when fire safety of the building may be affected.

ENVIRONMENT AGENCY (SCOTTISH ENVIRONMENTAL AGENCY)

■ responsible for authorising and regulating emissions from industry;

■ ensuring effective controls of the most polluting industries;

■ monitoring radioactive releases from nuclear sites;

■ ensuring that discharges to controlled waters are at acceptable levels;

■ setting standards and issuing permits for the collection, transporting, processing and disposal of waste (including radioactive waste);

■ enforcement of the Producer Responsibility Obligations (Packaging Waste) Regulations 1997; and

■ enforcement of the Waste Electrical and Electronic Equipment (WEEE) Directive and its associated directives.

INSURANCE COMPANIES

- legal requirement for employers to insure against liability for injury or disease to their employees arising out of their employment;

- offer fire and public liability insurance; and

- can influence health and safety standards by weighting the premium offered to an organisation against its health and safety record.

EMPLOYER DUTIES AND RESPONSIBILITIES

Under the HSW Act, the employer has a duty to safeguard the health, safety and welfare at work of:

- employees;

- other workers – agency, temporary or casual;

- trainees;

- contractors;

- visitors; and

- neighbours and the general public.

KEY ACTIONS REQUIRED OF THE EMPLOYER

- ensure the availability of competent advice on health and safety matters;

- obtain current Employers' Liability insurance and display the certificate;

- compile a health and safety policy and ensure that an adequate health and safety management system is in place;

- ensure that risk assessments of all work activities are undertaken and any required controls are put in place;

- provide the workforce with health and safety information and training;

- provide adequate welfare facilities;

- consult the workforce on health and safety issues;

- report and investigate some accidents, diseases and dangerous occurrences; and

- display prominently the health and safety law poster (or supply workers with the appropriate leaflet).

OTHER EMPLOYER RESPONSIBILITIES

Visitors and the general public

Possible hazards:

- unfamiliarity with the workplace processes;

- lack of knowledge of the workplace layout;

- unfamiliarity with emergency procedures;

- inappropriate personal protective equipment;

- inadequate or unsigned walkways; and

- added vulnerability if young or disabled visitors.

Possible controls:

- visitor identification (use of badges);

- routine signing in and out;

- escorted by a member of staff;

- provision of information on hazards and emergency procedures;

- given explicit site rules (wearing of PPE); and

- clear marking of walkways.

For night workers, employers should:

- determine the normal working time for night workers;

- if the working time is more than 8 hours per day on average, determine whether the amount of hours can be reduced;

- develop an appropriate health assessment and method of making health checks;

- ensure that proper records of night workers are maintained, including details of health assessments; and

- ensure that night workers are not involved in work which is particularly hazardous.

DUTIES AND RESPONSIBILITIES OF EMPLOYEES AND OTHERS

EMPLOYEES AND AGENCY WORKERS

Under the HSW Act, employees and agency workers have specific duties:

- take care for the health and safety of themselves and others who may be affected by their acts or omissions;

- co-operate with their employer, as far as is necessary, to ensure compliance with any statutory health and safety duty; and

- not to interfere with or deliberately misuse anything provided, in accordance with health and safety legislation, to further health and safety at work.

SELF-EMPLOYED

Under the HSW Act the self-employed are:

- responsible for their own health and safety; and

- responsible for ensuring that others who may be affected by their under-taking are not exposed to risks to their health and safety.

PERSONS IN CONTROL OF PREMISES

The duty of 'Persons in control of **non-domestic** premises' under Section 4 of the HSW extends to:

- people entering the premises to work

- people entering the premises to use machinery or equipment
- access to and exit from the premises
- corridors, stairs, lifts and storage areas.

DIRECTORS' RESPONSIBILITIES

Directors and board members should ensure that:

- health and safety arrangements are properly resourced;
- competent health and safety advice is obtained;
- regular reports are received on health and safety performance;
- any new or amended health and safety legislation is implemented;
- risk assessments are undertaken;
- there are regular audits of health and safety management systems and risk control measures; and
- there is adequate consultation with employees on health and safety issues.

Managing directors / chief executives, line managers and supervisors play key roles in ensuring that the health and safety policy is delivered and monitored.

DUTIES AND RESPONSIBILITIES OF MANUFACTURERS AND OTHERS IN THE SUPPLY CHAIN

Section 6 of the HSW Act (covered earlier) requires persons who design, manufacture, import or supply any article or substance for use at work to ensure, so far as is reasonably practicable, that it is safe and without risk to health, and employers should be provided with information on the safe use, dismantling and disposal of the articles and substances and given revised information should a subsequent serious risk become known.

Importers have a duty to ensure articles or substances comply with the requirements of UK legislation.

ADVANTAGES OF GOOD SUPPLY CHAIN MANAGEMENT

- reduction of waste of materials and time;

- ability to respond rapidly to changing requirements at short notice;

- reduction in accidents due to close liaison;

- legal duties addressed satisfactorily; and

- easier to maintain good customer care.

MANAGEMENT OF HEALTH AND SAFETY AT WORK REGULATIONS

EMPLOYERS' DUTIES

- undertake suitable and sufficient written risk assessments (five or more employees);

- put in place effective health and safety management arrangements and record them if five or more employees;

- employ a competent health and safety person;

- develop suitable emergency arrangements and inform employees and others;

- provide health and safety information to employees and others, such as other employers, the self-employed and their employees who are sharing the same workplace, and parents of child employees or those on work experience;

- co-operate in health and safety matters with other employers who share the same workplace;

- provide employees with adequate and relevant health and safety training;

- provide temporary workers and their contract agency with appropriate health and safety information;

- protect new and expectant mothers and young persons from particular risks; and

- under certain circumstances, as outlined in Regulation 6, provide health surveillance for employees.

The information that should be supplied by employers under the Regulations is:

- risks identified by any risk assessments, including those notified to them by other employers sharing the same workplace;

- the preventative and protective measures that are in place; and

- the emergency arrangements and procedures and the names of those responsible for the implementation of the procedures.

EMPLOYEES' DUTIES

- use any equipment or substance in accordance with any training or instruction given by the employer;

- report to the employer any serious or imminent danger; and

- report any shortcomings in the employer's protective health and safety arrangements.

SELF-EMPLOYED DUTIES

- carry out appropriate risk assessment;

- co-operate with other people who work in the premises and, where necessary, in the appointment of a health and safety co-ordinator; and

- provide comprehensible information to other people's employees working in their premises.

JOINT OCCUPATION OF PREMISES

The Regulations specifically state that where two or more employers share a workplace – whether on a temporary or a permanent basis – each employer shall:

- co-operate with other employers;

- take reasonable steps to co-ordinate between other employers to comply with legal requirements; and

- take reasonable steps to inform other employers where there are risks to health and safety.

RESPONSIBILITIES OF CLIENTS AND THEIR CONTRACTORS

Contractors are protected by the HSW Act (Section 3). The following points should be considered when contractors are employed:

- Health and safety must be included in the contract specification.

- All significant hazards must be included in the contract specification.

- The contractor must be selected with safety in mind.

- Prior to the start of work, health and safety policies should be exchanged.

- The contractor must be given basic site and health and safety information, such as welfare and first aid arrangements, significant hazards, safe storage of chemicals and the name of the contract supervisor.

- Where appropriate, the contractor should supply risk assessments and method statements.

- For construction work, the contractor should be aware of the position of buried/overhead services and the arrangements for the disposal of waste.

- The contractor should be monitored during the progress of the contract by the contract supervisor.

- The contract supervisor should check that the work has been completed safely at the end of the contract.

SAFETY RULES FOR CONTRACTORS

Contractors' safety rules should contain as a minimum the following points:

- **Health and safety** – that the contractor operates to at least the minimum legal standard and conforms to accepted industry good practice;

- **Supervision** – that the contractor provides a good standard of supervision of their own employees;

- **Sub-contractors** – that they may not use sub-contractors without prior written agreement from the organisation; and

- **Authorisation** – that each employee must carry an authorisation card issued by the organisation at all times while on site.

SELECTION OF CONTRACTORS

The following issues should be considered:

- an adequate health and safety policy;

- competent supervision;

- the availability of competent health and safety advice;

- past accident record;

- ability to assess hazards and risks involved in the contract and implement appropriate control measures;

- examples of method statements;

- a competent, trained and experienced workforce; and

- good references from previous contracts.

THE MANAGEMENT OF CONSTRUCTION ACTIVITIES (CDM REGULATIONS)

The duty holders and their main duties, when the work is notifiable, are:

- The **Client** has a major role and must ensure that:

  - all duty holders are competent

  - pre-construction information is passed to relevant duty holders

  - a CDM Co-ordinator is appointed

  - a Principal Contractor is appointed

  - a health and safety plan is developed by the Principal Contractor

  - a health and safety file is prepared; and

  - adequate resources are available for the safe completion of the work.

■ The **Designer** must:

- ensure that adequate health and safety provision is incorporated into the design

- provide adequate information about any significant risks associated with the design; and

- co-ordinate their work with that of the other duty holders.

■ The **CDM Co-ordinator** must:

- inform HSE of notifiable projects

- provide a link between duty holders, particularly with designers

- collect the pre-construction information and advise clients on any gaps

- manage the flow of information between clients, designers and contractors

- advise the client on the suitability of initial construction phase health and safety plan; and

- ensure that health and safety file is developed.

■ The **Principal Contractor** must:

- develop and manage the health and safety construction phase plan

- ensure that the construction phase is properly planned, managed and monitored

- employ only competent sub-contractors and supply them with necessary health and safety information

- ensure that suitable welfare facilities are provided on site; and

- display the project notification.

REQUIRED DOCUMENTATION

1 **Pre-construction information** – needs to identify the hazards and risks associated with the design and construction work.
2 **Health and safety file** – is a record of health and safety information required by the subsequent users of the finished construction project.
3 **Construction health and safety plan** – defines the organisation and arrangements required to control the site risks and co-ordinate the construction work.
4 **Notifiable work notice** – issued to the HSE if the construction work is to last longer than 30 days or involves more than 500 person days of work.

Health and Safety Management Systems – Policy

LEARNING OUTCOMES

Outline the key elements of a health and safety management system ■

Explain the purpose and importance of setting policy for health and safety ■

Describe the key features and appropriate content of an effective health and safety policy ■

KEY REVISION POINTS

The six key elements of the health and safety management
system HSG 65 ☐

The legal requirements for a health and safety policy ☐

The three key elements of a health and safety policy ☐

Target setting for health and safety performance ☐

Circumstances leading to the need for a review of health
and safety policy ☐

The reasons for unsuccessful health and safety policies ☐

KEY ELEMENTS OF A HEALTH AND SAFETY
MANAGEMENT SYSTEM (HSG 65)

The six elements of HSG 65 are:

1 **Policy** – A clear health and safety policy contributes to business efficiency
and continuous improvement throughout the operation. The policy should
state the intentions of the organisation in terms of clear aims, objectives,
targets and senior management involvement.

2 **Organising** – A well-defined health and safety organisation should identify
health and safety responsibilities at all levels of the organisation. An
effective organisation will be noted for good communication, the promotion
of competency, the commitment of all employees and a responsive reporting
system.

3 **Planning and implementing** – A clear health and safety plan based on risk
assessment and sets and implements performance standards, targets and
procedures through an effective health and safety management system. The
plan should set priorities and objectives for the control or elimination of
hazards and the reduction of risks.

4 **Measuring performance** – This includes both active (sometimes called
proactive) and reactive monitoring of the health and safety management

system. It is also important to measure the organisation against its own long-term goals and objectives.

5 **Reviewing performance** – The results of monitoring and independent audits should indicate whether the objectives and targets set in the health and safety policy need to be changed. Changes in the health and safety environment in the organisation, such as an accident, should also trigger a performance review. Performance reviews should include comparisons with internal performance indicators and the external performance indicators of similar organisations with exemplary practices and high standards.

6 **Auditing** – An independent and structured audit of all parts of the health and safety management system reinforces the review process. If the audit is to be really effective, it must assess both the compliance with stated procedures and the performance in the workplace. It will identify weaknesses in the health and safety policy and procedures and identify unrealistic or inadequate standards and targets.

Figure 3 The framework for Health and Safety Management HSG 65

PURPOSE AND IMPORTANCE OF SETTING POLICY FOR HEALTH AND SAFETY

The purpose of setting a health and safety policy is to ensure that:

- everybody associated with the organisation is aware of its health and safety aims and objectives and how they are to be achieved;

- the performance of the organisation is enhanced in areas other than health and safety;

- there is effective personal development of the workforce;

- business efficiency is improved throughout the operation; and

- the involvement of senior management in health and safety issues is evident to all stakeholders.

The policy should state the intentions of the business in terms of clear aims, objectives, organisation, arrangements and targets for all health and safety issues.

The law requires that the written health and safety policy should include the following three sections:

- a health and safety **policy statement of intent** which includes the health and safety aims and objectives of the organisation;

- the health and safety **organisation** detailing the people with specific health and safety responsibilities and their duties; and

- the health and safety **arrangements** in place in terms of systems and procedures.

LEGAL REQUIREMENTS

Section 2 of the HSW Act and the Employers' Health and Safety Policy Statements (Exception) Regulations 1975 require employers with **five or more** employees to prepare and review on a regular basis a written health and safety policy, together with the necessary organisation and arrangements to carry it out and to bring the policy and any revision of it to the notice of their employees.

Policy statement of intent	Organisation	Arrangements to include
• Aims and objectives	• Safety manual	• Planning and organising
• Duties of employer and employees	• Organisational chart	• Accident reporting
• Performance targets and benchmarks	• Responsibilities	• Emergencies
• Name of person responsible for health and safety	• Allocation of resources including finance	• Contractors and visitors
• Posted and dated	• Safety monitoring system	• Consultation and communication with employees
	• Identification of main hazards	• Fire precautions
		• Main risk assessments and hazard control
		• Performance monitoring

REASONS FOR A REVIEW OF THE HEALTH AND SAFETY POLICY

■ Significant organisational and/or technological changes have taken place.

■ There have been changes in personnel and/or legislation.

■ Health and safety performance has fallen below the occupational group's benchmarks.

■ The monitoring of risk assessments and/or accident/incident investigations indicates that the health and safety policy is no longer totally effective.

■ Enforcement action has been taken by the HSE or Local Authority.

■ A sufficient period of time has elapsed since the previous review.

THE EFFECTS OF A POSITIVE HEALTH AND SAFETY PERFORMANCE

are to:

- support the overall development of personnel;

- improve communication and consultation throughout the organisation;

- minimise financial losses due to accidents and ill-health and other incidents;

- directly involve senior managers in all levels of the organisation;

- improve supervision, particularly for young persons and those on occupational training courses;

- improve production processes; and

- improve the public image of the organisation or company.

THE REASONS FOR UNSUCCESSFUL HEALTH AND SAFETY POLICIES

include:

- the statements in the policy and the health and safety priorities are not understood by, or properly communicated to, the workforce;

- minimal resources are made available for the implementation of the policy;

- too much emphasis on rules for employees and too little on management policy;

- a lack of parity with other activities of the organisation (such as finance);

- lack of senior management involvement in health and safety;

- inadequate personal protective equipment;

- unsafe and poorly maintained machinery and equipment; and

- a lack of health and safety monitoring procedures.

Health and Safety Management Systems – Organising

1.3

LEARNING OUTCOMES

Outline the organisational health and safety roles and responsibilities of employers, directors and managers ■

Explain the concept of health and safety culture and its significance in the management of health and safety in an organisation ■

Outline the human factors which influence behaviour at work in a way that can affect health and safety ■

Explain how health and safety behaviour at work can be improved ■

Outline the need for emergency procedures and the arrangements for contacting emergency services ■

Outline the requirements for, and effective provision of, first aid in the workplace ■

🔑 KEY REVISION POINTS

The health and safety roles and responsibilities of employers
to their employees and others affected by their undertaking ☐

The health and safety roles and responsibilities of directors,
managers and supervisors ☐

The competence and responsibilities of the health and
safety practitioner/adviser ☐

The definition and importance of a health and safety culture ☐

The relationship between culture and performance ☐

The definition of human factors and their influence on the culture ☐

The development of a positive health and safety culture ☐

The importance of good communication ☐

The different forms of health and safety training ☐

The internal and external influences on the health and
safety culture of an organisation ☐

The duties of employers to consult with the workforce ☐

The development of emergency and first aid procedures
in the workplace ☐

ORGANISATIONAL HEALTH AND SAFETY ROLES AND RESPONSIBILITIES OF EMPLOYERS, DIRECTORS AND MANAGERS

ORGANISATIONAL HEALTH AND SAFETY RESPONSIBILITIES – EMPLOYERS

The organisational health and safety responsibilities of employers are covered in detail in 1.1.

ORGANISATIONAL HEALTH AND SAFETY RESPONSIBILITIES – DIRECTORS

Effective health and safety performance comes from the top, and directors have both collective and individual responsibility for health and safety.

Directors and Board members must ensure that	Management of health and safety at Board level involves
• the health and safety of employees and others, such as members of the public, is protected; • risk management includes health and safety risks and becomes a key business risk in board decisions; and • health and safety duties imposed by legislation are followed.	• planning the direction of health and safety; • delivering the plan for health and safety; • monitoring health and safety performance; and • reviewing health and safety performance.

ORGANISATIONAL HEALTH AND SAFETY RESPONSIBILITIES – MANAGERS

1. Managing Directors / Chief Executives

Managing Directors / Chief Executives are responsible for:

■ the health and safety performance within the organisation;

■ ensuring that adequate resources are available for the health and safety requirements within the organisation;

■ the establishment, implementation and maintenance of a health and safety programme for the organisation that encompasses all areas of significant health and safety risk;

■ the approval, introduction and monitoring of all site health and safety policies, rules and procedures; and

■ the review and possible revision annually of the effectiveness of the health and safety programme.

2. Departmental managers

The principal departmental managers may report to the Site Manager, Managing Director or Chief Executive. In particular, they:

- are responsible and accountable for the health and safety performance of their department;

- must ensure that any machinery, equipment or vehicles used within the department are maintained, correctly guarded and meet agreed health and safety standards. Copies of records of all maintenance, statutory and insurance inspections must be kept by the Departmental Manager;

- develop a training plan that includes specific job instructions for new or transferred employees and follow up on the training by supervisors. Copies of records of all training must be kept by the Departmental Manager;

- personally investigate all lost workday cases and dangerous occurrences and report to their line manager. Progress any required corrective action.

3. Supervisors

The supervisors are responsible to and report to their Departmental Manager. In particular, they:

- are responsible and accountable for their team's health and safety performance;

- enforce all safe systems of work procedures that have been issued by the Departmental Manager;

- instruct employees in relevant health and safety rules, make records of this instruction and enforce all health and safety rules and procedures;

- enforce personal protective equipment requirements, check that it is being used and periodically appraise condition of equipment; and

- record any infringements of the personal protective equipment policy.

HEALTH AND SAFETY ADVISER

Must:

- be competent following the attainment of a health and safety qualification and training;

- report directly to senior management on matters of policy;

- keep up to date with technological advances and legislative changes;

- advise on the establishment of health and safety, maintenance and accident investigation procedures; and

- provide liaison with external agencies, such as the HSE, Fire Authorities, contractors, insurance companies and the public.

CONCEPT OF HEALTH AND SAFETY CULTURE

The safety culture of an organisation is the product of individual and group values, attitudes, perceptions, competencies and patterns of behaviour that determine the commitment to, and the style and proficiency of, an organisation's health and safety management.

FEATURES OF A GOOD HEALTH AND SAFETY CULTURE

- leadership and commitment to health and safety at all levels;

- acceptance that high standards are achievable;

- mutual trust throughout the organisation;

- detailed risk assessments and control and monitoring procedures;

- health and safety policy, including a code of practice and required health and safety standards;

- training, communication and consultation systems;

- encouragement to the workforce to report potential hazards;

- health and safety monitoring systems; and

- prompt accident investigation and implementation of remedial actions.

INDICATORS OF A HEALTH AND SAFETY CULTURE

■ accident/incident rates;

■ sickness and absentee rates;

■ resources available for health and safety;

■ level of legal and other compliance;

■ turnover rates for employees;

■ level of complaints;

■ selection and management of contractors;

■ levels and effectiveness of communication and supervision;

■ health and safety management structure; and

■ level of insurance premiums.

HUMAN FACTORS

1 in 10 near misses = 1 accident.

90% accidents due to human error – 70% poor management.

Human factors are affected by the:

■ Organisation

■ Job

■ Personal factors.

ORGANISATION

■ must have a positive health and safety culture;

■ manage health and safety by providing leadership and involvement of senior managers;

■ motivate the workforce to improve health and safety performance; and

■ measure health and safety performance.

JOB

- recognise possibility of human error;
- good ergonomics, equipment design and layout of workstation;
- clear job descriptions;
- safe systems of work and operating procedures;
- job rotation and regular breaks;
- provision of correct tools; and
- effective training schedule and good communication.

PERSONAL FACTORS

The three common psychological factors are:

- Attitude – tendency to behave in a particular way in a given situation, influenced by social background and peer pressure
- Motivation – the driving force behind the way a person acts or is stimulated to act
- Perception – the way in which a person believes or understands information supplied or a situation.

Other related factors:

- self-interest – e.g. effect of bonus systems;
- position in the team;
- acknowledgement by management of good work/initiatives;
- hearing and/or memory loss;
- experience and competence;
- age, personality, attitude, language problems;
- training undertaken and information given;
- effect of shift working – e.g. night working; and
- health (physical and mental).

Human errors **may be:**	Violations **may be:**
1. Slips – failure to carry out the correct actions of a task	1. Routine – the breaking of a safety rule or procedure is the normal way of working
2. Lapses – failure to carry out particular actions that form part of a working procedure	2. Situational – job pressures at a particular time make rule compliance difficult
3. Mistakes are: • rule-based – a rule or procedure is applied or remembered incorrectly or • knowledge-based – well tried methods or calculation rules are applied incorrectly	3. Exceptional – a safety rule is broken to perform a new task

METHODS OF IMPROVING HEALTH AND SAFETY BEHAVIOUR AT WORK

DEVELOPMENT OF A POSITIVE HEALTH AND SAFETY CULTURE

By:

1 Commitment of management is the most important factor:

- Proactive management

- Promotion by example (e.g. wearing PPE)

2 The promotion of health and safety standards for:

- Selection and design of premises

- Selection and design of plant, processes and substances

- Recruitment of employees and contractors

- Risk assessments and control implementation

- Competence, maintenance and supervision

- Emergency planning and training

- Transportation of the product and its subsequent maintenance and servicing

3   Competence of the workforce, including in health and safety:

- Knowledge and understanding of the work/job

- Capacity to apply knowledge to the particular task

- Awareness of one's limitations.

COMMUNICATION

- Verbal (by mouth) – conversations, telephone

- Written – memos, emails, meeting minutes, data sheets

- Graphic – safety signs, posters, charts.

Use of notice boards

The approved poster entitled '*Health and Safety Law – what you should know*' must be displayed on a notice board or the approved leaflet distributed.

In addition to the health and safety poster, the following types of health and safety information could be displayed on a workplace notice board:

- a copy of the Employers' Liability Insurance Certificate;

- details of first aid arrangements;

- emergency evacuation and fire procedures;

- minutes of the last health and safety committee meeting;

- details of health and safety targets and performance against them; and

- health and safety posters and campaign details.

Barriers to effective communication

- language and dialect

- acronyms and jargon

- various physical and mental disabilities

- attitudes and perception of workers and supervisors.

Types of accident propaganda

- Statistics
- Films, DVDs and posters
- Targets
- Records.

For safety propaganda to be effective, it must have:

- a simple understandable message
- a positive believable message
- an appealing format that will motivate the reader.

CONSULTATION WITH EMPLOYEES

Safety representatives from a recognised trade union have the following rights:

- to investigate accidents and dangerous occurrences;
- to investigate health and safety complaints;
- to undertake workplace inspections;
- to receive information from health and safety inspectors;
- to attend health and safety committee meetings;
- to have access to suitable facilities to perform their functions; and
- to be allowed time off with pay for health and safety training.

If two or more representatives request in writing for a health and safety committee to be set up then the employer must comply within 3 months.

SAFETY COMMITTEES

Terms of reference should include the following:

- the study of accident and notifiable disease statistics to enable reports to be made of recommended remedial actions;

- the examination of health and safety audits and statutory inspection reports;

- the consideration of reports from the external enforcement agency;

- the review of new legislation, Approved Codes of Practice and guidance and their effect on the organisation;

- the monitoring and review of all health and safety training and instruction activities in the organisation;

- the monitoring and review of health and safety publicity and communication throughout the organisation;

- the development of safe systems of work and safety procedures;

- the reviewing risk assessments;

- the considering reports from safety representatives; and

- the continuous monitoring of arrangements for health and safety and revising them whenever necessary.

Representatives of Employee Safety (ROES) were established under the Consultation with Employees Regulations.

The employer must **inform** and **consult** employees on health and safety matters.

ROES have the following functions:

- inform the employer of health and safety concerns of the workforce;

- inform the employer of potential hazards and dangerous occurrences in the workplace;

- inform the employer of any general matters that affect the health and safety of the workforce; and

- speak on behalf of the workforce to health and safety inspectors.

Employer must consult on:

■ new processes or equipment or any changes in them;

■ the appointment arrangements for a health and safety competent person;

■ the results of any risk assessments;

■ the arrangements for the management of health and safety training; and

■ the introduction of new technologies.

Types of information that employer does NOT need to disclose if it:

■ violates a legal prohibition

■ endangers national security

■ relates to a specific individual

■ could harm the company commercially

■ was obtained from legal proceedings.

TYPES OF HEALTH AND SAFETY TRAINING

■ Induction – at recruitment

■ Job-specific

■ Supervisory and management

■ Specialist (e.g. first aid)

■ Refresher or reinforcement.

Internal influences on health and safety culture	External influences on health and safety culture
Management commitment	Expectations of society
Production/service demands	Legislation and enforcement
Communication	Insurance companies
Competence	Trade unions
Employee representation	State of the economy
	Commercial stakeholders

EMERGENCY PROCEDURES

EXAMPLES OF TYPES OF EMERGENCIES

- fire
- explosions, bomb scares
- escape of toxic gases
- major accident.

TYPICAL ELEMENTS OF EMERGENCY PROCEDURES

- Fire notices and fire procedures (including testing)
- Fire drills and evacuation procedures
- Assembly and roll call
- Arrangements for contacting emergency and rescue services
- Provision of information for emergency services
- Internal emergency organisation – including control of spillages and clean-up arrangements
- Media and publicity arrangements
- Business continuity arrangements.

PROVISION OF FIRST AID

MAIN FUNCTIONS OF FIRST AID TREATMENT

- preservation of life and/or minimisation of the consequences of serious injury until medical help is available
- treatment of minor injuries not needing medical attention.

MAIN FIRST AID REQUIREMENTS

- qualified first aiders
- adequate facilities and equipment to administer first aid
- an assessment of required first aid cover and requirements
- an appointed person available to assist first aiders.

BASIC FIRST AID PROVISION (INCLUDING NUMBER OF FIRST AIDERS)

Depends on:

- number of workers;
- the hazards and risks in the workplace;
- accident record and types of injuries;
- proximity to emergency medical services; and
- working patterns (shift work).

First aider requires first aid training and the first aid course depends on the level of risk in the workplace.

- For low risk, a 6-hour Emergency First Aid at Work course
- For higher risk, 18-hour First Aid at Work course
- Either 6-hour emergency or 4-day first aid course initially
- A repeat course every 3 years.

Appointed person has some other first aid experience/qualification.

First aid box – contents depend on particular workplace needs but should be checked regularly.

Health and Safety Management Systems – Planning

1.4

LEARNING OUTCOMES

Explain the importance of planning in the context of health and safety management systems ☐

Explain the principles and practice of risk assessment ☐

Explain the general principles of control and a basic hierarchy of risk reduction measures ☐

Identify the key sources of health and safety information ☐

Explain what factors should be considered when developing and implementing a safe system of work ☐

Explain the role and function of a permit-to-work system ☐

KEY REVISION POINTS

The legal requirements for a health and safety risk assessment ☐

The meaning of 'suitable and sufficient' ☐

The forms and objectives of risk assessment ☐

Details of accident categories and types of health risks ☐

The risk assessment process and its management ☐

Groups that require a special risk assessment ☐

The principles of prevention and the hierarchy of risk control ☐

The development and application of safe systems of work
and permits to work for various activities, including those
in confined spaces ☐

Hazards and controls for confined spaces and machinery
maintenance work ☐

THE IMPORTANCE OF PLANNING

The health and safety planning process begins with finding:

- the correct information about the existing management system for health and safety;

- suitable benchmarks against which to make comparisons; and

- competent people to carry out the analysis and make sensible judgements.

Further judgement may be necessary to establish if the system is:

- adequate for the organisation and the range of hazards/risks;

- working as intended and achieving the right objectives; and

- delivering cost-effective and proportionate risk control in the workplace.

SETTING HEALTH AND SAFETY OBJECTIVES

The standard of the health and safety objectives of an organisation will depend upon:

- the seniority of the person setting the objectives;

- the formal documentation associated with the objectives and their relevance at each functional level in the organisation;

- the incorporation of relevant legal, technological options and good practice guidance;

- the identification and assessment of all significant hazards and risks within the organisation;

- the success with which they integrate with financial, operational and business requirements; and

- the incorporation of the views of employees, stakeholders and other interested parties.

The objectives must be **specific**, **measurable**, **achievable** and with **realistic timescales (SMART)** and kept up to date with any changes in legislation.

PRINCIPLES AND PRACTICE OF RISK ASSESSMENT

LEGAL REQUIREMENT

Regulation 3 of Management of Health and Safety at Work Regulations requires written suitable and sufficient risk assessment if there are five or more employees.

Suitable and sufficient means:

- identify significant risks only;

- identify measures required to comply with legislation; and

- remain appropriate and valid over a reasonable period of time.

Hazard – the potential to cause harm

Risk – the likelihood to cause harm

Residual risk – the risk remaining after some controls are in place.

FORMS OF RISK ASSESSMENT

- ■ Quantitative – calculated from risk = likelihood × severity

- ■ Qualitative – descriptor (high, medium or low) used to describe timetable for remedial action

- ■ Generic – covers similar activities or work equipment.

HEALTH RISKS

- ■ Chemical – exhaust fumes, paint solvents, asbestos

- ■ Biological – legionella, pathogens, hepatitis

- ■ Physical – noise, vibration, radiation

- ■ Psychological – stress, violence

- ■ (Ergonomic – musculoskeletal disorders).

RISK ASSESSMENT PROCESS

- ■ Hazard identification – Step 1 of HSE's five steps

- ■ Persons at risk – Step 2 of HSE's five steps

  - ● employees, agency/temporary workers, contractors, shift workers

  - ● members of the public – visitors, customers, patients, students, children, elderly

  - ● special groups – young persons, expectant or nursing mothers, workers with a disability, lone workers.

- ■ Evaluation of risk level (residual risk) – Step 3 of HSE's five steps

  - ● high, medium and low (defined qualitatively or quantitatively)

  - ● both occupational and organisational risk levels need to be considered.

- ■ Detail risk controls (existing and additional) – Step 3 of HSE's five steps

  - ● the prioritisation of risk control is important

- risks can be reduced at the design stage by using the principles of prevention

- risks can be controlled by using the hierarchy of risk control (see Element 6).

■ Record of risk assessment findings – Step 4 of HSE's five steps

■ Monitor and review – Step 5 of HSE's five steps.

Regular reviews required but need to be more frequent if:

- new legislation introduced

- new information available on substances or process

- changes to the workforce – introduction of trainees

- an accident has occurred.

RISK ASSESSMENT TEAM REQUIREMENTS

■ all need training in risk assessment

■ leader should have health and safety experience

■ all need to be competent to assess risks in area under examination

■ all need to know their own limitations

■ include local line manager in the team

■ at least one team member with report writing skills.

SPECIAL CASES

Young persons

■ are under 18 years

■ covered by Regulation 19 MHSWR

■ subject to peer pressure and are inexperienced

■ are eager to please

■ appropriate level and approach of subject matter in training sessions.

There are additional requirements if under school leaving age. **A special risk assessment must be made to include details of**:

■ the work activity;

■ any prohibited processes or equipment;

■ the health and safety training provided; and

■ the supervision arrangements.

Expectant and nursing mothers

There are restrictions on the type of work that can be undertaken.

Risks include:

■ manual handling;

■ chemical and biological agents;

■ ionising radiation;

■ passive smoking;

■ lack of rest room facilities;

■ temperature variations;

■ prolonged standing or sitting; and

■ stress and violence to staff.

Workers with a disability

■ emergency arrangements including 'raising the alarm'; and

■ adequate wheelchair access to fire exit.

Lone workers

- special risk assessment (including violence);
- must be fit to work alone;
- special training should be given;
- possible to handle all equipment and substances alone;
- periodic visits by supervisor;
- regular mobile phone contact with base;
- first aid arrangements; and
- emergency arrangements.

GENERAL PRINCIPLES OF CONTROL

The planning and implementing section of the health and safety management system is based on risk assessment and concerns all the actions taken to control or eliminate hazards and reduce risks. The principles of prevention are used when equipment or processes are being designed or selected. The hierarchy of risk control enables risk to be further controlled.

THE PRINCIPLES OF PREVENTION

- avoid risks
- evaluate risks which cannot be avoided
- adapt work to the individual
- adapt to technical changes
- replace dangerous items with less dangerous items
- develop an overall prevention policy
- give priority to collective measures (Safe Place strategy)
- give instructions to employees (Safe Person strategy).

HIERARCHY OF RISK CONTROL

1 elimination
2 substitution
3 changing work methods/patterns
4 reduced time exposure
5 engineering controls (isolation, insulation and ventilation)
6 good housekeeping
7 safe systems of work
8 training and information
9 personal protective equipment
10 welfare
11 monitoring and supervision
12 review.

SOURCES OF INFORMATION ON HEALTH AND SAFETY

Internal sources	External sources
• accident and ill-health records and investigation reports • absentee records • inspection and audit reports undertaken by the organisation and by external organisations such as the HSE • maintenance, risk assessment (including COSHH) and training records	• health and safety legislation • HSE publications, such as approved codes of practice, guidance documents, leaflets, journals, books and their website • international (e.g. ILO), European and British standards • health and safety magazines and journals
• documents which provide information to workers • any equipment examination or test reports	• information published by trade associations, employer organisations and trade unions • specialist technical and legal publications • information and data from manufacturers and suppliers • the internet and encyclopaedias

SAFETY SIGNS

- Red – prohibition – round (e.g. no smoking)

- Yellow – warning – triangular (e.g. wet floor)

- Blue – mandatory – round (e.g. ear defenders must be worn)

- Green – safe condition – square or rectangular (e.g. first aid).

DEVELOPMENT AND IMPLEMENTATION OF SAFE SYSTEMS OF WORK

EMPLOYER'S DUTY

This is contained in the common law duty of care and in Section 2 of the Health and Safety at Work Act.

ROLE OF COMPETENT PERSONS

A competent person or safety adviser should:

- assist managers to draw up guidelines for safe systems of work;

- prepare suitable documentation; and

- advise management on the adequacy of the safe systems produced.

ROLE OF MANAGERS

- provide safe systems of work;

- ensure that employees are adequately trained in a specific safe system of work and are competent to carry out the work safely; and

- provide sufficient supervision to ensure that the system of work is followed and the work is carried out safely.

EMPLOYEE INVOLVEMENT

Includes:

- consultation with those employees who will be exposed to the risks, either directly or through their representatives;

- discussion of the proposed system with those who will have to work under it and supervise it; and

- understanding that employees have a responsibility to follow the safe system of work.

STEPS IN THE DEVELOPMENT OF A SAFE SYSTEM OF WORK

1 Assess the task (complexity, accident records etc.)
2 Identify the significant hazards associated with the task
3 Define safe methods for performing the task (including emergency procedures) – document the methods if required
4 Implement the safe system of work (written safe system to be signed off)
5 Monitor the safe system of work and review it if necessary
6 Train the workforce in safe procedures and enter on training record.

Safe systems of work are particularly important for:

- maintenance work

- contractors

- emergency procedures

- lone working

- vehicle operations

- cleaning operations.

Method statements are formal written safe systems of work and are often used in construction work.

The safe system of work should be based on a thorough analysis of the job or operation to be covered by the system. After the introduction of a safe system of work the following controls are required:

■ Engineering or process controls (e.g. guarding)

■ Documented procedures

■ Behavioural controls requiring a certain standard of behaviour from individuals.

Procedures for lone workers may include:

■ periodic visits from the supervisor;

■ regular voice contact;

■ automatic warning devices to alert others of problems;

■ checks that the lone worker has returned safely;

■ special arrangements for first aid to deal with minor injuries; and

■ emergency arrangements.

ROLE AND FUNCTION OF A PERMIT-TO-WORK

Permits to work are:

■ **formal** safe systems of work;

■ required to be signed on/off by a responsible person; and

■ often require equipment to be locked on/off by a responsible person.

To be used whenever there is a high risk of serious injury, such as:

■ confined spaces

■ live electricity (particularly high voltage)

■ hot working (e.g. welding work)

■ some machinery maintenance work.

Typical responsible persons are:

- site manager
- senior authorised person – often the chief engineer
- authorised person – issues permit
- competent person – receives permit
- operatives – supervised by a competent person
- specialists (e.g. electrical engineer)
- engineers – usually responsible for the work
- contractors.

CONFINED SPACES

Examples include underground chamber, silo, trench, sewer, tunnel.

Hazards are:

- lack of oxygen and asphyxiation
- poor ventilation
- presence of fumes
- poor means of access and escape
- drowning
- claustrophobia
- electrical equipment (needs to be flameproof)
- presence of dust (e.g. silos)
- heat and high temperatures
- fire and/or explosion
- poor or artificial lighting.

Controls include:

- permit-to-work

- risk assessment

- training and information for all workers entering the confined space

- emergency arrangements in place

- emergency training

- no entry for unauthorised persons.

MACHINERY MAINTENANCE WORK

Hazards are:

- no perceived risk

- no safe system of work

- poor communications

- failure to brief contractors

- lack of familiarity

- poor design.

Control of hazards requires:

- effective planning

- a written safe system of work/permit-to-work

- a risk assessment to assess, control and reduce risks

- monitoring to ensure that the system of work and controls are used

- an effective training programme for all involved in the work.

Health and Safety Management Systems – Measuring, Audit and Review

1.5

LEARNING OUTCOMES

Outline the principles, purpose and role of active and reactive monitoring

Explain the purpose of, and procedures for, health and safety auditing

Explain the purpose of, and procedures for, investigating incidents (accidents, cases of work-related ill health and other occurrences)

Describe the legal and organisational requirements for recording and reporting incidents

Explain the purpose of, and procedures for, regular reviews of health and safety performance

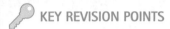 **KEY REVISION POINTS**

The reasons for measuring and monitoring health and safety performance ☐

The role of standards in the monitoring process ☐

The differences between reactive and active (proactive) monitoring ☐

The importance of clear, unambiguous report writing ☐

The role of performance review and audit ☐

The advantages and disadvantages of internal and external audit ☐

The reasons and legal requirements for recording and reporting incidents and accidents ☐

Basic accident investigation procedures and the different types of accident and incident ☐

Immediate and underlying causes of incidents ☐

Issues concerning insurance and compensation claims ☐

WORKPLACE INSPECTIONS

Detailed check, often using a checklist, of the whole workplace and should cover:

■ the premises (e.g. fire precautions, access/egress, housekeeping);

■ the plant, equipment and substances (e.g. machine guarding, tools, ventilation);

■ the procedures in place (e.g. safe systems of work, risk assessments, use of PPE); and

■ the workforce (e.g. training, information, supervision, health surveillance).

ACTIVE AND REACTIVE MONITORING

Reactive monitoring (taking action after a problem occurs) involves	Proactive (or active) monitoring (taking action before problems occur) involves
• review of accident and ill-health reports – often to check that remedial advice has been actioned or ascertain trends and hot spots • review of procedures following dangerous occurrences, other property damage and near misses • review of compensation claims • review of complaints from the workforce and members of the public	• the active monitoring of the workplace for unsafe conditions • the direct observation of workers for unsafe acts • meeting with management and workers to discover any problems • checking documents, such as maintenance records, near miss reports, insurance reports
• review of procedures following enforcement reports and notices • review of risk assessments following the discovery of additional hazards	• undertaking workplace inspections, sampling, surveys, tours and audits

Other issues with inspections

- the competence of the observers;

- the frequency of inspections;

- the response to the inspection report; and

- the use of objective inspection standards.

Safety sampling – checking for safety defects in a selected area of the workplace (e.g. all fire extinguishers).

Safety surveys – a detailed inspection of a particular workplace activity throughout an organisation (e.g. manual handling).

Safety tours – an unscheduled brief inspection of a work area in the workplace by a team led by a senior manager.

HEALTH AND SAFETY AUDITING

Auditing – the independent collection of information on the efficiency, effectiveness and reliability of the whole health and safety management system measured against specific standards. It will check that the following are in place:

■ appropriate management arrangements;

■ adequate risk control systems exist and implemented;

■ appropriate workplace precautions; and

■ appropriate documentation and records.

Audits should take place at regular intervals.

OTHER ISSUES WITH AUDITS

External audits:

■ are independent of organisation

■ are competent

■ are familiar with external benchmarks

■ are more expensive than internal audits

■ can offer bland reports.

Internal audits:

■ are not independent of organisation and partiality may be questioned

■ usually require audit training

■ are less expensive than external audits

■ know the organisation well, particularly critical areas

■ can spread good practice around the organisation

■ may be unaware of external benchmarks.

INVESTIGATING INCIDENTS

Purpose of accident/incident investigation is to:

- eliminate the cause and future occurrences;

- determine the direct and indirect causes of the accident/incident;

- define any corrective and/or preventative actions;

- identify any deficiencies in risk controls, the health and safety management system and/or procedures;

- ensure that all legal requirements are being met; and

- comply with the recommendations of the Woolf Report so that essential information is available in the event of a civil claim.

Benefits of accident/incident investigation:

- prevention of a recurrence;

- prevention of future business losses;

- prevention of future increased insurance premiums and costs of criminal and civil actions; and

- improve employee morale and organisation reputation.

ACCIDENT CAUSATION

Direct or immediate include	Indirect, root or underlying include
• unsafe acts by individuals due to poor behaviour or a lack of training, supervision, information or competence or a failure to wear PPE • unsafe conditions, such as inadequate guarding, inadequate procedures, hazardous substances, ergonomic and/or environmental factors	• poor machine maintenance and/or start-up procedures • management and social pressures • financial restrictions • lack of management commitment to health and safety • poor or lack of health and safety policy and standards

(Continued overleaf)

Direct or immediate include	Indirect, root or underlying include
• fire or explosive hazards	• poor workplace health and safety culture • workplace and trade customs and attitudes

ELEMENTS OF AN INVESTIGATION

■ interview relevant operatives, managers, supervisors

■ obtain detailed plans and/or photographs of scene

■ check all relevant records (working procedures, maintenance, training, risk assessments)

■ interview all witnesses

■ possibly arrange for equipment and/or substances to be independently tested

■ produce a concise report for management that includes

  ● background to the accident

  ● possible causes of the accident (direct and indirect)

  ● relevant health and safety legislation, guidance and standards

  ● recommendations (including any remedial actions)

  ● any additional training or follow-up requirements

■ undertake a post-accident risk assessment.

RECORDING AND REPORTING INCIDENTS

ACCIDENT REPORTING REQUIREMENTS

■ the organisational accident/incident report form

■ accident book

■ RIDDOR 95 – records kept for 3 years.

RIDDOR 95

The Regulations apply to employees, the self-employed, trainees and contractors. Any fatality or injury to a member of the public resulting in off-site medical treatment is reportable.

Fatality	• reportable by quickest means (telephone) • submit form 2508 within 10 days • reportable if death occurs within 1 year of accident
Major accident	• reportable as soon as possible • submit form 2508 within 10 days • examples – any fracture, other than fingers, thumbs or toes; any amputation: any other injury requiring admittance to hospital for more than 24 hours
Over 7 (was 3) days' lost time injury	• submit form 2508 within 10 days
Disease	• forthwith • submit form 2508A within 10 days • examples – hepatitis; occupational dermatitis; hand–arm vibration syndrome
Dangerous occurrence	a 'near miss' that could lead to serious injury or loss of life. • reportable by quickest means (telephone) • submit form 2508 within 10 days • examples – collapse of scaffolding; the collapse, overturning or failure of any load-bearing part of lifts and lifting equipment; explosion or fire resulting in the suspension of normal working for more than 24 hours

COMPENSATION AND INSURANCE ISSUES

■ importance of documentation and records

■ need for pre- and post-accident risk assessments

■ impact of Woolf Report and documentation required to defend a civil claim.

REVIEW OF HEALTH AND SAFETY PERFORMANCE

Performance review is the final stage of the management process and reviews:

■ all monitoring, inspection and audit reports;

■ the adequacy of the health and safety management system, and performance standards against external benchmarks;

■ whether new legislation or guidance has been applied;

■ whether the health and safety policy objectives have been met or need modification to ensure continuous improvement; and

■ whether there has been adequate feedback to/from managers.

Reviews should aim to include:

■ evaluation of compliance with legal and organisational requirements;

■ incident data, recommendations and action plans from investigations;

■ inspections, surveys, tours and sampling;

■ absences and sickness records and their analysis;

■ any reports on quality assurance or environmental protection;

■ audit results and implementation;

■ monitoring of data, reports and records;

■ communications from enforcing authorities and insurers;

■ any developments in legal requirements or best practice within the industry;

■ changed circumstances or processes;

■ benchmarking with other similar organisations;

■ complaints from neighbours, customers and the public;

■ effectiveness of consultation and internal communications;

■ whether health and safety objectives have been met; and

■ whether actions from previous reviews have been completed.

The Board should review health and safety performance at least once a year. The review process should:

■ examine whether the health and safety policy requires revision;

■ examine whether risk management and other health and safety issues have been effectively reported to the Board;

■ report health and safety shortcomings;

■ address any weaknesses with effective measures and develop a system to monitor their implementation; and

■ undertake immediate reviews in the light of major shortcomings or events.

Controlling Workplace Hazards

Workplace Hazards and Risk Control

LEARNING OUTCOMES

Outline common health, welfare and work environment
requirements in the workplace

Explain the risk factors and appropriate controls for violence
at work

Explain the effects of substance misuse on health and
safety at work and control measures to reduce such risks

Explain the hazards and control measures for the safe
movement of people in the workplace

Explain the hazards and control measures for safe
working at height

Explain the hazards of, and control measures for, excavations

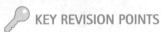 KEY REVISION POINTS

Welfare and work environment requirements ☐

The causes and prevention of workplace violence ☐

The hazards and control strategies for pedestrian safety ☐

The hazards and controls involved with working at height, including roof work ☐

Access equipment, including scaffolds, ladders and MEWPs ☐

Emergency rescue procedures ☐

The hazards and controls required for excavation work ☐

HEALTH, WELFARE AND WORK ENVIRONMENT REQUIREMENTS

Welfare and work environment issues are covered by the Workplace (Health, Safety and Welfare) Regulations.

WELFARE

◼ sanitary conveniences and washing facilities

◼ drinking water

◼ clothing and changing facilities

◼ rest and eating facilities.

WORKPLACE ENVIRONMENT

◼ ventilation

◼ heating and temperature (particularly extremes)

- lighting
    - natural light
    - stroboscopic effects
    - special local lighting
    - structural aspects (shadows)
    - atmospheric dust
    - heating effect of lighting
    - lamp and window cleaning
    - emergency lighting
- workstations and seating
- floors, stairways and traffic routes
- translucent or transparent doors should be constructed with safety glass and properly marked to warn pedestrians of their presence
- adequate arrangements in place to ensure the safe cleaning of windows and skylights
- adequate provisions for the needs of disabled workers.

EFFECTS OF EXTREME TEMPERATURE

	Health effects	Preventative measures
High temperatures	Heart strain Heat stroke	Supply of drinking water Supply of ventilated air Regular health surveillance Information and training on the hazards
Low temperatures	Frostbite Loss of limbs	Supply of thick, warm (thermal) clothing Provision of hot drinks and external heating Regular health surveillance Information and training on the hazards

VIOLENCE AT WORK

Defined as any incident in which a person is abused, threatened or assaulted in circumstances related to their work. High-risk occupations include health and social services, police and fire fighters, lone and night workers and benefit services.

ACTION PLAN

1 find out if there is a problem
2 decide on what action to take:

- quality of service provided
- design of the operating environment
- type of equipment used
- designing the job

3 take the appropriate action
4 check that the action is effective.

TYPE OF SECURITY EQUIPMENT USED

- access control (swipe cards and simple coded security locks)
- closed circuit television
- alarms – there are three main types:

- intruder alarms
- panic alarms
- personal alarms

- radios and pagers (particularly for lone workers)
- mobile phones.

SUBSTANCE MISUSE AT WORK

RISKS TO HEALTH AND SAFETY OF DRUGS AND ALCOHOL

- reduce productivity;
- increase absenteeism;
- increase accidents at work; and, in some cases,
- endanger the public.

Alcohol abuse is a considerable problem when vehicle driving is part of the job, especially if driving is required on public roads.

CONTROLS

For alcohol abuse:

- during working hours, there should be no drinking
- drinking during break periods should be discouraged
- induction training should stress this policy
- managers should enforce the policy
- posters can also help to communicate the message
- the availability of a confidential counselling service.

Drug abuse presents similar problems to those found with alcohol abuse:

- absenteeism
- reduced productivity
- an increase in the risk of accidents
- impairments in cognition, perception and motor skills.

SIGNS OF A SUBSTANCE MISUSE PROBLEM

- sudden mood changes
- unusual irritability or aggression

- a tendency to become confused

- abnormal fluctuations in concentration and energy

- impaired job performance

- poor time-keeping

- increased short-term sickness leave

- a deterioration in relationships with colleagues, customers or management

- dishonesty and theft (arising from the need to maintain an expensive habit).

A policy on drug abuse can be established by:

1 Investigation of the size of the problem
2 **Planning actions** – develop an awareness programme for all staff and a special training programme for managers and supervisors
3 **Taking action** by producing a written policy to include:

- details of the safeguards to employees

- the confidentiality given to anyone with a drug problem

- the circumstances in which disciplinary and/or reported action will be taken

4 **Monitoring the policy** by checking for positive changes in the measures made during the initial investigation (improvements in the rates of sickness and accidents).

SAFE MOVEMENT OF PEOPLE IN THE WORKPLACE

Hazards to pedestrians include:

- slips, trips and falls on the same level

- falls from height

- collisions with moving vehicles

- being struck by moving, falling or flying objects

- striking against fixed or stationary objects.

Trip hazards are caused by	Slip hazards are caused by
• loose floorboards or carpets • obstructions, low walls, low fixtures on the floor • cables or trailing leads to portable electrical hand tools and other electrical appliances across walkways • raised telephone and electrical sockets • rugs and mats – particularly when worn or placed on a polished surface • poor housekeeping – obstacles left on walkways, rubbish not removed regularly • poor lighting levels – particularly near steps or other changes in level • sloping or uneven floors – particularly where there is poor lighting or no handrails • unsuitable footwear – shoes with a slippery sole or lack of ankle support	• wet or dusty floors • the spillage of wet or dry substances – oil, water, plastic pellets • loose mats on slippery floors • wet and/or icy weather conditions • unsuitable footwear, floor coverings or sloping floors

CONTROL STRATEGIES FOR PEDESTRIANS

■ risk assessment to identify suitable controls

■ slip-resistant surfaces and reflective edges to stairs and kerbs

■ spillage control and drainage

■ designated walkways

■ fencing and guarding, particularly on stairways

■ use of warning signs

■ sound storage racking that is inspected and maintained regularly

■ maintenance of a safe workplace

- cleaning and good housekeeping procedures

- access and egress

- environmental considerations (heating, lighting, noise and dust)

- special provision for disabled people

- high visibility clothing, appropriate footwear, personal protective equipment (safety harnesses)

- information, instruction, training and supervision.

WORKING AT HEIGHT

The Work at Height Regulations have no minimum height requirement for work at height. They include all work activities where there is a need to control a risk of falling a distance liable to cause personal injury. They would therefore include:

- working on a scaffold or from a mobile elevated work platform (MEWP);

- sheeting a lorry or dipping a road tanker;

- tree surgery and other forestry work at height;

- using cradles or rope for access to a building or other structure such as a ship under repair;

- climbing permanent structures such as a gantry or telephone pole;

- working near an excavation area or cellar opening if a person could fall into it and be injured;

- painting or pasting and erecting bill posters at height;

- work on staging or trestles (e.g. for filming or events);

- using a ladder/stepladder or kick stool for shelf filling, window cleaning and the like; and

- working in a mine shaft or chimney.

THE REGULATIONS GIVE THE FOLLOWING HIERARCHY OF CONTROL

1 avoid working at height, if possible;
2 the provision of a properly constructed working platform, complete with toe boards and guard rails;
3 if this is not practicable or where the work is of short duration, suspension equipment should be used; and only when this is impracticable
4 collective fall arrest equipment (air bags or safety nets) may be used;
5 where this is not practicable, individual fall restrainers (safety harnesses) should be used; and
6 only when none of the other measures are practicable should ladders or stepladders be considered.

FRAGILE ROOFS AND SURFACES

A roof should always be treated as fragile until a competent person has ruled otherwise. Fragility can be caused by:

■ general deterioration of the roof through ageing, neglect and lack of maintenance

■ corrosion of cladding and fixings

■ quality of the original installation and selection of materials

■ thermal and impact damage

■ deterioration of the supporting structure

■ weather damage.

Asbestos cement sheets and old roof lights should always be treated as fragile.

Other roof work hazards

■ overhead services and obstructions

■ the presence of asbestos or other hazardous substances

■ the use of equipment such as gas cylinders and bitumen boilers

■ manual handling hazards.

REQUIREMENTS FOR HEAD PROTECTION

The Construction (Head Protection) Regulations virtually mandate employers to supply head protection (hard hats) to employees whenever there is a risk of head injury from falling objects. Other requirements:

■ the employer must ensure that hard hats are properly maintained and replaced when they are damaged;

■ self-employed workers must supply and maintain their own head protection;

■ visitors to construction sites should always be supplied with head protection; and

■ mandatory head protection signs displayed around the site.

These regulations will soon be included in the PPE Regulations.

FALL ARREST EQUIPMENT

The three most common types are:

■ safety harnesses

■ safety nets

■ air bags.

EMERGENCY PROCEDURES (INCLUDING RESCUE)

A suitable emergency and rescue procedure needs to be in place for situations that could be expected where people are working at height.

Factors to be considered:

■ the method of rescue must be simple and straightforward;

■ any rescue system selected needs to be proportionate to the risk;

■ there should not be undue reliance on the emergency services. However, arrangements should be in place to inform the emergency services of any serious incident;

■ awareness of suspension trauma;

- rescue teams should be available at all times with a designated leader; and

- instruction and training are essential for the teams, preferably by simulated rescue exercises.

ACCESS EQUIPMENT

Ladders

The following factors should be considered when using ladders:

- aluminium ladders are light but should not be used in high winds or near live electricity

- timber ladders need regular inspection for damage and should not be painted as this could hide cracks

- ensure that the use of a ladder is the safest means of access for the work to be done and the height to be climbed

- the ladder needs to be stable in use with a safe inclination (1 in 4)

- the foot of the ladder should be tied to a rigid support

- the proximity of live electricity should be checked

- there should be at least 1 metre of ladder above the stepping-off point

- over-reaching must be eliminated

- workers who are to use ladders must be trained in the correct method of use and selection

- ladders should be inspected (particularly for damaged or missing rungs) and maintained on a regular basis and they should only be repaired by competent persons.

Certain work should not be attempted using ladders. This includes work where:

- two hands are required

- the work is at an excessive height

- the ladder cannot be secured or made stable

- the work is of long duration

- the work area is very large
- the equipment or materials to be used are heavy or bulky
- the weather conditions are adverse
- there is no protection from vehicles.

Fixed scaffold

Fixed scaffolds are usually independently tied and important considerations are:

- scaffolding must only be erected and dismantled by competent people – any changes to the scaffold must be made by a competent person;
- adequate toe boards, guard rails and intermediate rails must be fitted to prevent people or materials from falling;
- the scaffold must rest on a stable surface; uprights should have base plates and timber sole plates if necessary;
- the scaffold must have safe access and egress;
- work platforms should be fully boarded with no tipping or tripping hazards;
- the scaffold should be sited away from or protected from traffic routes so that it is not damaged by vehicles;
- the scaffold should be properly braced and secured to the building or structure;
- overloading of the scaffold must be avoided;
- the public must be protected at all stages of the work; and
- regular inspections of the scaffold must be made and recorded.

Components of a scaffold

- Standard
- Ledger
- Guard rail
- Toe boards

- Bracing
- Transom
- Base plate
- Sole board
- Ties
- Working platform

Prefabricated mobile scaffold towers

The following points must be considered:

- the selection, erection and dismantling of mobile scaffold towers must be undertaken by competent persons;
- maximum height to base ratios must not be exceeded;
- diagonal bracing and stabilisers should be used;
- access ladders must be fitted to the narrowest side of the tower or inside the tower;
- persons should not climb up the frame of the tower;
- all wheels must be locked whilst work is in progress;
- all persons must vacate the tower before it is moved;
- the tower working platform must be boarded, fitted with guard rails and toe boards and not overloaded;
- towers must be tied to a rigid structure if exposed to windy weather;
- persons working from a tower must not over-reach or use ladders from the work platform;
- safe distances must be maintained between the tower and overhead power lines both during working operations and when the tower is moved; and
- the tower should be inspected on a regular basis and a report made.

MOBILE ELEVATED WORK PLATFORMS (MEWPS)

The following factors must be considered when using mobile elevated work platforms:

■ the mobile elevated work platform must only be operated by trained and competent persons;

■ safety harnesses should be worn;

■ it must never be moved in the elevated position;

■ it must be operated on level and stable ground;

■ the tyres must be properly inflated and the wheels immobilised;

■ outriggers should be fully extended and locked in position;

■ due care must be exercised with overhead power supplies and obstructions; and

■ procedures should be in place in the event of machine failure.

ROOF WORK

The main hazards are:

■ fragile roofing materials – more brittle with age and exposure to sunlight

■ exposed edges

■ unsafe access equipment

■ fragile roof-lights and voids

■ falls from girders, ridges or purlins

■ overhead services and obstructions

■ the use of equipment such as gas cylinders and bitumen boilers

■ manual handling hazards.

The controls include:

- a risk assessment and a method statement must be completed before any roof work takes place;

- suitable means of access such as scaffolding, ladders and crawling boards;

- suitable barriers, guard rails or covers where people work near to fragile materials and roof lights;

- edge protection on a flat roof is best provided by double guard rails and toe board;

- suitable warning signs indicating that a roof is fragile, at ground level;

- access to the area immediately below the work should be restricted using suitable barriers, netting, safety signs and safety helmets;

- the safe transportation of materials to and from the working area, involving lifts, hoists, manual handling and/or using chutes and covered waste skips; and

- good housekeeping procedures.

A risk assessment for roof work should include:

- an assessment of the structural integrity of the roof

- the methods to be used to repair the roof

- a COSHH assessment of any hazardous substances to be used.

CAUSES OF ACCIDENTS WITH ACCESS EQUIPMENT

- Design faults

- Over-reaching

- Falls and slips

- Collision with vehicles

- Unsound base support

- Untrained personnel

- Climbing with loads
- Poor maintenance and/or inspection
- Instability
- Adverse wind effects.

INSPECTION AND MAINTENANCE

Equipment for work at height needs regular inspection and maintenance. The requirements for inspection are:

- the name and address of the person for whom the inspection was carried out
- the location of the work equipment inspected
- a description of the work equipment inspected
- the date and time of the inspection
- details of any matter identified that could give rise to a risk to the health or safety of any person
- details of any action taken as a result of any matter identified above
- details of any further action considered necessary
- the name and position of the person making the report.

EXCAVATIONS

HAZARDS

- Collapse of sides
- Materials falling on workers
- Falls of people and/or vehicles into excavation
- Influx of groundwater
- Underground services

■ Access and egress

■ Proximity of waste or stored materials or adjacent structures

■ Fumes and health hazards (Weil's disease).

PRECAUTIONS

■ Hard hats

■ Shore sides

■ Barrier around top

■ Well lit at night

■ Identify position of buried services

■ Care during the filling in

■ Supervision by a competent person.

INSPECTIONS

■ Every 7 days with a report

■ Every day –

● After any event likely to affect the stability of excavation

● Before start of each shift

● After an accidental fall of material.

Transport Hazards and Risk Control

LEARNING OUTCOMES

Explain the hazards and control measures for the safe movement of vehicles in the workplace

Outline the factors associated with driving at work that increase the risk of an incident and the control measures to reduce work-related driving risks

KEY REVISION POINTS

The hazards and control strategy for safe vehicular operations	☐
The management of vehicle movements in the workplace	☐
The management of occupational driving risks in terms of the driver, vehicle and journey planning	☐
The particular issues in the use of lorries	☐

SAFE MOVEMENT OF VEHICLES IN THE WORKPLACE

Hazards in vehicle operations include:

- collisions between pedestrians and vehicles

- vehicles crushing feet of pedestrians

- people falling from vehicles

- people being struck by objects falling from or attached to vehicles

- people being struck by or ejected from an overturning vehicle

- communication problems between vehicle drivers and employees or members of the public.

Hazards in vehicle operations may be caused by:

- poor working practices, such as the lack of regular vehicular safety checks

- defective maintenance, steering, brakes, tyres and hydraulic hoses

- poor road surfaces and/or poorly drained road surfaces

- overloading of vehicles

- use of unsuitable vehicles to transport people

- inadequate supervision

- poor training or lack of refresher training.

Other more general hazards involving pedestrians and vehicles include:

- reversing of vehicles, especially inside buildings
- roadways too narrow with insufficient safe parking areas
- roadways poorly marked out and inappropriate or unfamiliar signs used
- too few pedestrian crossing points
- the non-separation of pedestrians and vehicles
- lack of barriers along roadways
- lack of directional and other signs
- poor visibility due to load impeding view, blind corners or inadequate vehicular lighting or mirrors
- poor environmental factors, such as lighting, dust and noise
- ill-defined speed limits and/or speed limits which are not enforced
- vehicles used by untrained and/or unauthorized personnel.

HAZARDS FROM MOBILE WORK EQUIPMENT

Accidents involving mobile work equipment often arise from one or more of the following events:

- poor maintenance with defective brakes, tyres and steering
- poor visibility
- operating on rough ground or steep gradients
- carrying of passengers without the proper accommodation for them
- people being ejected as the vehicle overturns and/or being crushed by it
- being struck by a vehicle or an attachment
- lack of driver training or experience
- poor management procedures, training and supervision
- collision with other vehicles

- overloading of vehicles

- general vehicle movements and parking

- dangerous occurrences or other emergency incidents

- access and egress from the buildings.

CONTROL STRATEGIES FOR SAFE VEHICLE OPERATIONS

- risk assessment

- designated marked traffic routes with good visibility and well-designed loading and storage areas

- enforced speed limits and/or the fitting of speed governors

- audible warning of approach signals, particularly for reversing

- separation of pedestrians and vehicles

- use of one-way systems

- the issue of suitable high visibility clothing and other personal protective equipment to drivers and pedestrians in vehicle operating areas

- separate site access gates for pedestrians and vehicles

- use of pedestrian crossings on vehicular routes

- mirrors on blind corners

- identification of recognised parking and non-parking areas

- include vehicle safety in employee induction training

- driver training and refresher training given by competent persons

- the provision of driver protection by the use of:

  - seat belts

  - FOPS (fall over protection system)

  - ROPS or TOPS (roll or tip over protection systems)

  - banksmen during reversing operations

- effective vehicular maintenance procedures

- use of daily vehicular inspection checklists

- environmental considerations (road surfaces, gradients and changes in road level, lighting, visibility)

The management of vehicle movements involves:

- a designated management system and a code of practice for all drivers, who must be competent

- an effective line management organisation to ensure supervision of all vehicular activities

- a documented preventative maintenance programme with all work recorded

- regular inspections and, in some cases, thorough examinations by competent persons

- ensure that all vehicles are fitted with suitable driver protection – seat belts, ROPS

- ensure that all vehicles are fitted with reversing warning systems

- suitable fire precautions, particularly in battery charging areas.

DRIVING AT WORK

FACTORS AFFECTING OCCUPATIONAL ROAD RISKS

The driver:

- competency issues including:

  - validity of the driving licence

  - awareness of the company policy on work-related road safety

  - general driving experience

- health and fitness, in particular eyesight

- level of training received.

The vehicle:

- its suitability;

- its condition, particularly seat belts and head restraints;

- its past maintenance schedule;

- the ergonomics of the driving position;

- any load should be stacked safely so that it cannot move during the journey. There must also be satisfactory arrangements for handling the load at either end of the journey; and

- the vehicle breakdown procedure.

The journey:

- route to be undertaken, including the amount of traffic, the time of day when the journey is made and the sufficiency of the time allowed for its completion;

- the type of roads being used, whether motorway, minor roads or those involving city centre congestion, and their condition;

- schedules need to take sufficient account of busy periods on roads;

- driving excessive distances without appropriate breaks; and

- adverse weather conditions, such as snow, ice, heavy rain and high winds.

MANAGING OCCUPATIONAL ROAD RISKS

An effective policy to manage occupational road risk should address important and relevant aspects of the Highway Code, the possible enforcement action by the authorities and the following issues:

For the driver

- all drivers must have a current and valid driving licence;

- details of a prospective driver's previous driving experience should be examined and recorded;

- driver training and health;

- regular eyesight tests;

- staff should not drive, or undertake other duties, while taking a course of medicine that might impair their judgement;

- some prescribed and over the counter drugs and medicines can affect driver awareness and speed of reaction. Drivers using such medicines must check with a doctor or pharmacist to ensure that it is safe to drive;

- the monitoring of driver performance;

- the number of recorded incidents;

- all speed limits must be observed;

- the determination and recording of driving hours;

- routine driver safety checks such as those on lights, tyres and wheel fixings;

- drivers and passengers are adequately protected in the event of an incident. Crash helmets and protective clothing for those who ride motorcycles and other two-wheeled vehicles should be of the appropriate colour and standard;

- drivers of Heavy Goods Vehicles (HGVs) must have the appropriate driving licence and medical certificate; and

- disciplinary action taken in the event of non-compliance with the agreed in-house rules and standards.

For the vehicle

- all vehicles must carry comprehensive insurance for use at work;

- all vehicles should be fit for purpose;

- new or replacement vehicles should be suitable for both driving and the health and safety of the public;

- all vehicles should be regularly serviced and maintained – particularly important when pool cars are used, because pool car users often assume another user is checking on maintenance and the MOT;

■ the safety critical systems (brakes, steering and tyres) need to be properly maintained;

■ if the vehicle is leased and serviced by the leasing company, a system should be in place to confirm that servicing is being done to a reasonable standard;

■ each vehicle should have a handbook available for the driver;

■ there should be a defect reporting system for drivers to use if they consider their vehicle is unsafe; and

■ if privately owned vehicles are used for work, they should be insured for business use and, if necessary, have an appropriate MOT certificate test.

For lorries

■ the policy should address issues associated with the loads;

■ if the load is hazardous, emergency procedures (and possibly equipment) must be in place and the driver trained in those procedures;

■ the load should be stacked safely in the lorry so that it cannot move during the journey;

■ there must also be satisfactory arrangements for handling the load at either end of the journey;

■ the driver must be aware of the height of the vehicle, both laden and empty; and

■ where tachographs are carried, they should be checked regularly to make sure that drivers are not driving for long periods without a break.

For journey planning

■ journey times must include statutory breaks and allow for possible delays;

■ motorways are the safest roads – minor roads are suitable for cars, but are less safe and could present difficulties for larger vehicles;

■ overhead restrictions, for example bridges or tunnels, and other hazards such as level crossings may present dangers for long and/or high vehicles;

- sufficient consideration will need to be given to adverse weather conditions; and

- no alcohol or recreational drugs must be consumed during the day of the journey until the journey is completed. Only minimal amounts of alcohol should be consumed on the day before a journey is to be made.

Musculoskeletal Hazards and Risk Control

2.3

LEARNING OUTCOMES

Explain work processes and practices that may give rise to work-related upper limb disorders and appropriate control measures ▪

Explain the hazards and control measures that should be considered when assessing risks from manual handling activities ▪

Explain the hazards, precautions and procedures to reduce the risk in the use of lifting and moving equipment, with specific reference to manually operated load moving equipment ▪

Explain the hazards and the precautions and procedures to reduce the risk in the use of lifting and moving equipment, with specific reference to mechanically operated load moving equipment ▪

🔑 KEY REVISION POINTS

Musculoskeletal health issues and ill–health prevention strategies ☐

The types of injuries resulting from manual handling operations ☐

The development of manual handling risk assessments and effective control measures ☐

Safety in the use of lifting equipment and an understanding of the different types of manual and mechanical handling and lifting equipment ☐

Hazards and controls associated with fork lift trucks, hoists and cranes ☐

The legal requirements for the inspection and examination of lifting equipment ☐

WORK–RELATED UPPER LIMB DISORDERS

Ergonomics – interaction between the worker, his work and his environment. It involves knowledge of his physical and mental capabilities in addition to his understanding of the job.

Musculoskeletal disorders (MSDs) affect muscles, joints, tendons, ligaments and nerves. MSDs usually affect the back, neck, shoulders and upper limbs (termed work-related upper limb disorders – WRULDs).

ILL–HEALTH EFFECTS DUE TO POOR ERGONOMICS

Work-related upper limb disorders (WRULDs):

■ Tenosynovitis

■ Repetitive strain injury (RSI)

■ Carpal tunnel syndrome

■ Frozen shoulder.

Often caused by repetitive operations – keyboard operations, brick laying, assembly of small components.

DISPLAY SCREEN EQUIPMENT (DSE)

Requirements of the Health and Safety (Display Screen Equipment) Regulations:

- Suitable and sufficient risk assessment
- Workstation compliance with minimum specifications given in the Regulations
- Adequate breaks in work programme
- Free sight test if required
- Information and training.

Ill-health hazards from DSE work:

- Musculoskeletal problems
- Visual problems
- Psychological problems.

DSE risk assessment

The DSE risk assessment should consider the following factors:

- the height and adjustability of the monitor
- the adjustability of the keyboard, the suitability of the mouse and the provision of wrist support
- the stability and adjustability of the DSE user's chair
- the provision of ample foot room and suitable foot support
- the effect of any lighting and window glare at the workstation
- the storage of materials around the workstation
- the safety of trailing cables, plugs and sockets
- environmental issues – noise, heating, humidity and draughts.

MANUAL HANDLING HAZARDS, RISKS AND CONTROL MEASURES

Manual handling can involve any load movement by human effort only (lifting, pushing, pulling, carrying or supporting).

REQUIREMENTS OF THE MANUAL HANDLING OPERATIONS REGULATIONS

- to avoid manual handling, if possible;

- to mechanise or automate the lifting process, if possible;

- if unavoidable, undertake a risk assessment;

- instigate risk controls and review; and

- give employees information on loads.

Manual handling hazards:

- lifting a load which is too heavy and/or cumbersome

- poor posture and/or poor lifting technique

- dropping the load on the foot

- lifting sharp-edged or hot loads.

Injuries caused by manual handling include:

- muscular sprains or strains

- back injuries

- trapped nerve

- hernias

- cuts, bruises and abrasions

- fractures

- work related upper limb disorders (WRULDs)

- rheumatism.

Manual handling risk assessment requires the use of mechanical aids; if this is not possible, then the assessment of:

■ the task

■ the individual

■ the load

■ the working environment.

Manual handling training includes:

■ the types of injury;

■ the manual handling assessment findings;

■ potentially hazardous manual handling operations;

■ the correct use of manually operated load moving equipment, such as sack and pallet trucks;

■ the correct use of mechanical aids;

■ the correct use of personal protective equipment;

■ good housekeeping issues;

■ the factors that can affect an individual; and

■ a good lifting technique, such as:

   1 Check that suitable clothing (including gloves, if required) and footwear are being worn. Make an approximate assessment of the weight of the load and decide whether to lift it alone. Check to see whether one side of the load is heavier than the other and ensure that the heaviest side of the load is closest to the body.

   2 Place the feet apart and adopt a good posture by bending the knees and keeping the back straight.

   3 Get a firm grip and hold the load as close as possible to the body.

   4 Ensuring that the back remains straight, lift the load to knee level and then to waist level without jerking.

5   After ensuring that full visibility is available, move forward without twisting the trunk.

6   Set the load down either at waist level or by lowering it first to knee level and then to the floor.

7   Manoeuvre the load to its final position after it has been set down.

MANUALLY OPERATED LOAD HANDLING EQUIPMENT

TYPES OF MANUALLY OPERATED LOAD HANDLING EQUIPMENT

- simple tools
- wheelbarrows
- trucks and trolleys
- roller tracks and chutes
- pallet trucks
- conveyors
- various types of hoists that can be used to lift people as well as other loads.

HAZARDS ASSOCIATED WITH MANUALLY OPERATED LOAD HANDLING EQUIPMENT

- incorrect use, such as overloading or attempting to carry unstable loads
- manual handling hazards due to pushing or pulling a truck or trolley
- lack of maintenance
- injuries to pedestrians using the same walkway.

PRECAUTIONS WITH THE USE OF MANUALLY OPERATED LOAD HANDLING EQUIPMENT

- when handling aids are being selected, the subsequent user should be consulted whenever possible;

- moving and handling tasks are often made easier by good design;

- a safe system of work should be in place;

- all operators of mechanical handling equipment must be properly trained in its use and supervised while they are using it;

- sufficient room available to manoeuvre the equipment easily;

- adequate visibility and lighting available;

- the floor is in a stable condition;

- regular safety checks to identify any faults with the equipment;

- a regular maintenance schedule for the equipment;

- the proposed use will be within the safe working load of the equipment;

- the lifting equipment is CE-marked; and

- advice is sought from the suppliers/hirers on its suitability for the proposed task and any maintenance requirements.

MECHANICALLY OPERATED LOAD HANDLING EQUIPMENT

POSITIONING AND INSTALLATION OF LIFTING EQUIPMENT

Risks during lifting operations should be reduced by:

- avoiding the equipment or its load striking a person;

- avoiding a load drifting, falling freely or being released unintentionally;

- avoiding the need to lift loads over people;

- stopping safely in the event of a power failure; and

- where possible, enclose the path of the load with suitable and substantial interlocked gates.

THE ORGANISATION OF LIFTING OPERATIONS

Every lifting operation, which is lifting or lowering a load, shall be:

- ◼ properly planned by a competent person
- ◼ appropriately supervised
- ◼ performed in a safe manner.

SUMMARY OF THE REQUIREMENTS FOR LIFTING OPERATIONS

There are four general requirements for all lifting operations:

- ◼ strong, stable and suitable lifting equipment;
- ◼ the equipment should be positioned and installed correctly;
- ◼ the equipment should be visibly marked with the safe working load; and
- ◼ lifting operations must be planned, supervised and performed in a safe manner by competent people.

TYPES OF MECHANICAL HANDLING AND LIFTING EQUIPMENT

CONVEYOR

There are three types of conveyor – belt, roller and screw.

Typical hazards include:

- ◼ in-running nips
- ◼ entanglement
- ◼ loads falling from conveyor
- ◼ impact against overhead systems
- ◼ contact hazards – sharp edges
- ◼ manual handling hazards
- ◼ noise and vibration hazards.

ELEVATOR

An elevator transports goods between floors. The most common hazard is caused by loads falling from the elevator. Also there are manual handling hazards at either end of the elevator.

FORK LIFT TRUCKS

Hazards include:

- overturning
- overloading
- collisions with pedestrians or other vehicles or structures
- the silent operation of electrically powered fork lift trucks makes pedestrians unaware of their presence
- uneven road surface
- overhead obstructions
- loss of load
- failure of the hydraulic system
- inadequate maintenance
- lack of driver training
- use as a work platform and/or carrying passengers
- speeding
- poor vision around load
- dangerous stacking technique, particularly on warehouse racking
- fire – when battery charging or refuelling.

There are also the following physical hazards:

- noise
- exhaust fumes

- vibrations

- manual handling

- ergonomic – musculoskeletal injuries due to soft tyres, expansion joints.

Driver daily checks:

- tyres and tyre pressures

- brakes

- reversing horn and light

- mirrors

- secure and properly adjusted seat

- correct fluid levels

- correct working of all lifting and tilting systems.

CRANES

PRINCIPLES OF SAFE OPERATION

- brief inspection prior to each use (including lifting tackle);

- loads not left suspended when crane not in use;

- prior to lift, ensure that nobody can be struck by load or crane;

- never carry loads over people;

- good visibility and communications;

- only lift loads vertically – do not drag load;

- travel with load as close to the ground as possible; and

- switch off power to crane when unattended.

MOBILE JIB CRANES

- plan each lift (maximum load and radius of operation);

- identify overhead obstructions and hazards;

- assess load-bearing capacity of ground; and

- if fitted, outriggers should be used.

REASONS FOR CRANE FAILURE

- overloading

- poor slinging of load

- overturning

- collision with another structure or overhead power lines

- foundation failure

- structural failure of the crane

- operator error

- lack of maintenance and/or regular inspections.

During lifting operations, ensure that:

- driver has good visibility;

- there are no pedestrians below load and barriers are in place; and

- audible warning given prior to lifting operation.

LIFTS, HOISTS AND ITEMS OF LIFTING TACKLE

A **lift** or **hoist** incorporates a platform or cage. Its movement is restricted by guides and it may carry passengers and/or goods alone.

They are required to:

- be of sound mechanical construction;

- have interlocking doors or gates that must be completely closed before the lift or hoist moves;

- be fitted, if carrying passengers, with an automatic braking system to prevent over-running and a safety device to support the lift in the event of suspension rope failure;

■ be rigorously maintained by competent persons; and

■ protect others during maintenance operations from falling down the lift shaft and other hazards.

Other **items of lifting tackle** include chain slings and hooks, wire and fibre rope slings, eyebolts and shackles.

Important points include:

■ loads must be properly secured and balanced in slings;

■ lifting hooks should be checked for wear and hook distortion;

■ shackles and eyebolts must be correctly tightened;

■ slings must be checked for any damage before use and only used by competent people;

■ training and instruction should be given in the use of lifting tackle;

■ regular inspections of tackle should be made in addition to the mandatory thorough examinations; and

■ lifting items should be carefully stored between uses.

The equipment should be inspected at suitable intervals between thorough examinations. The frequency and the extent of the inspection are determined by the level of risk presented by the lifting equipment. A report or record should be made of the inspection, which should be kept until the next inspection. Unless stated otherwise, lifts and hoists should be inspected every week.

STATUTORY EXAMINATION OF LIFTING EQUIPMENT

The Lifting Operations and Lifting Equipment Regulations (LOLER) require inspections and thorough examinations.

An **inspection** is performed by a competent person appointed by the employer and is used to identify whether the equipment can be operated, adjusted and maintained safely.

The following points apply to inspections:

■ take place between thorough examinations

■ the report or record is kept until next inspection

■ the frequency depends on risk assessment

■ lifts and hoists should be inspected every week.

A **thorough examination** is a detailed examination of lifting equipment by a competent person who is independent of the employer. The examination is usually made in accordance with a written scheme and a report (within 28 days) is submitted to the employer. The report must be kept for 2 years or until the next thorough examination (whichever is longer).

A thorough examination should be undertaken at the following times:

■ before first use;

■ after it has been assembled at a new location;

■ at least every 6 months for lifting equipment or lifting accessory used for lifting persons;

■ at least every 12 months for all other lifting equipment;

■ in accordance with a particular examination scheme drawn up by an independent competent person; and

■ every time exceptional circumstances, which are likely to jeopardise the safety of the lifting equipment, have occurred (such as severe weather).

Work Equipment Hazards and Risk Control

2.4

LEARNING OUTCOMES

Outline the general requirements for work equipment

Explain the hazards and controls for hand-held tools

Describe the main mechanical and non-mechanical hazards of machinery

Explain the main control measures for reducing risks from machinery hazards

KEY REVISION POINTS

The definition of work equipment ☐

The legal requirements for the supply and use of work equipment ☐

The safe use and maintenance of work equipment ☐

The hazards and controls related to hand-held tools ☐

Examples of mechanical and non-mechanical machinery hazards ☐

Hierarchy of measures to reduce the risks of injury from
work equipment ☐

Practical safeguards and their application to a range of machines ☐

The construction requirements of guards ☐

GENERAL REQUIREMENTS OF WORK EQUIPMENT

There are **two** types of law affecting work equipment:

■ equipment supply law – designed to prevent barriers to trade – The Supply
of Machinery (Safety) Regulations, and

■ equipment user law – designed to protect people at work – The Provision
and Use of Work Equipment Regulations.

SUPPLY OF WORK EQUIPMENT

Most new work equipment should have 'CE' marking when purchased. The 'CE'
mark requires manufacturers to:

■ identify the health and safety hazards likely to be present when the machine
is used;

■ assess the associated risks;

■ remove the significant hazards at the design stage or, if that is not
possible;

- provide safeguards (such as guarding or noise enclosures) or, if that is not possible;

- use warning signs on the machine to warn of hazards.

Additional requirements of equipment manufacturers:

- retain related information in a technical file;

- fix a CE mark to the equipment – indicates compliance with all relevant supply laws;

- issue a 'Declaration of Conformity' for the equipment.

The **Declaration of Conformity** states that the equipment complies with all relevant health and safety requirements. The Declaration must:

- state the name and address of the manufacturer or importer into the EU;

- contain a description of the equipment, its make, type and serial number;

- indicate all relevant European Directives with which the machinery complies;

- state details of any notified body that has been involved;

- specify which standards have been used in the manufacture (if any);

- be signed by a person with authority to do so; and

- provide the purchaser with instructions on the safe installation, use and maintenance of the equipment.

USE OF WORK EQUIPMENT

DEFINITION OF WORK EQUIPMENT

Any equipment used by an employee at work, such as:

- hand tools – hammers, knives, spanners

- machines – photocopiers, drilling machines, vehicles

- apparatus – laboratory equipment

- lifting equipment – fork lift trucks, hoists

- other equipment – ladders, vacuum cleaners

- installations – scaffolds, sound enclosures.

Uses of work equipment include:

- starting and stopping

- repair and maintenance

- modification and servicing

- cleaning

- transporting.

MANAGEMENT DUTIES

Work equipment must be:

- suitable for its purpose of use and only used for specified operations, and

- suitable in terms of initial integrity and place of use;

- selected with the health and safety of the user in mind;

- maintained in an efficient working order and in good repair;

- inspected at regular intervals;

- restricted in use to designated persons when it has specific hazards;

- used together with appropriate information, instruction and training; and

- in conformity with EC requirements.

Other duties include:

- the guarding of dangerous parts of work equipment;

- the provision of stop and emergency stop controls;

- ensuring that the equipment is stable;

- that the lighting around the equipment is suitable and sufficient; and

- suitable warning markings or devices, such as flashing lights, are fitted.

HIERARCHY OF RISK CONTROL FOR WORK EQUIPMENT

1 eliminate the risks
2 control risks by physical methods (such as guarding)
3 software measures (such as safe systems of work).

INFORMATION, INSTRUCTION AND TRAINING

This should include:

■ the significant health and safety issues with the equipment

■ any limitations or problems with the use of the equipment

■ safe methods to deal with these problems

■ details of any residual risks associated with the equipment

■ details of safe working procedures

■ the location of the manufacturer's manual or guidance.

Induction training will be required on recruitment, but refresher training will also be needed when:

■ jobs are changed, particularly if the level of risk increases;

■ if new technology or equipment is introduced;

■ the system of work is modified;

■ there are legislative changes; or

■ periodically as a refresher course.

TYPES OF MAINTENANCE

1 Preventative planned maintenance
2 Condition-based maintenance
3 Breakdown-based maintenance.

HAZARDS ASSOCIATED WITH MAINTENANCE

- lack of competence and/or training
- equipment not made safe before maintenance
- lack of permit-to-work
- use of incorrect tools or unsafe equipment.

INSPECTION OF WORK EQUIPMENT

- after installation or put into service for the first time;
- after assembly at a new site or in a new location; and thereafter
- at suitable intervals; and
- every time exceptional circumstances occur that could affect safety.

EQUIPMENT CONTROLS

The equipment should have:

- operating controls – easily reached and operated and with adequate markings and warning signs
- no obstructions or debris around it
- emergency controls (red emergency stop buttons)
- stability during use.

OTHER ISSUES

- the level and quality of general and local lighting
- clear and durable markings on equipment.

STATUTORY REQUIREMENTS FOR EXAMINATION OF BOILERS AND AIR RECEIVERS

- supplied with correct written information and markings;
- properly installed;
- used within its operating limits;
- written scheme for periodic examination certified by a competent person;
- examined by a competent person in accordance with the written scheme; and
- a report of the periodic examination held on file and any required actions undertaken.

RESPONSIBILITIES OF USERS

- take care for the health and safety of themselves and others who may be affected by their acts or omissions;
- co-operate with their employer, as far as is necessary, to ensure compliance with any statutory health and safety duty;
- not to interfere with or deliberately misuse anything provided, in accordance with health and safety legislation, to further health and safety at work;
- use any equipment or substance in accordance with any training or instruction given by the employer;
- report to the employer any serious or imminent danger; and
- report any shortcomings in the employer's protective health and safety arrangements.

HAND-HELD TOOLS

Typical hazards include:

- broken handles
- incorrect use of knives, saws and chisels
- tools that slip
- splayed spanners
- flying particles
- electrocution and/or burns.

Controls:

- suitable for purpose (e.g. insulated tools for electricians)
- inspection on a regular basis
- training for all hand tool users.

HAND-HELD POWER TOOLS

EXAMPLES OF HAND-HELD POWER TOOLS

- electric drills
- electric carving knives and food mixers
- disc-cutters/cut-off saws
- sanders
- chainsaws

Hazards include:

- entanglement
- flying particles (eye injury)
- contact with cutting edges
- striking hidden/buried services (electrical cables, gas mains)

- manual handling injuries

- hand–arm vibration syndrome

- noise levels

- dust levels

- trailing leads

- fire and/or explosion.

Typical controls:

- read operating instructions before initial use;

- maintain a clean, tidy and well-lit working area;

- never use power tools near water, combustible fluids or gases;

- store tools in a dry, safe and secure place;

- never overload tools;

- always use the correct tool and/or attachment for a heavy job;

- always wear suitable work clothes and appropriate personal protective equipment;

- ensure that the work piece is securely held in clamps or a vice;

- maintain tools with care, keep them clean and sharp and inspect cables regularly;

- always check that the power switch is turned off before connecting the power cable; and

- use only tool accessories and attachments recommended by the tool manufacturer.

The use of chainsaws is very hazardous because:

- there is no guard on the cutting chain

- they are heavy

- they are noisy.

MACHINERY HAZARDS

MECHANICAL MACHINERY HAZARDS

- crushing

- shearing

- cutting or severing

- entanglement

- drawing in or trapping (in-running nips)

- impact

- stabbing or puncture

- friction or abrasion

- high pressure fluid ejection.

NON-MECHANICAL MACHINERY HAZARDS INCLUDE

- access to the machinery

- slips, trips and falls

- falling and moving objects from the machinery

- obstructions and projections

- manual handling and lifting

- electricity (shock, burns or fire)

- fire and explosion

- noise and vibration

- high pressure ejection from hydraulic hose leaks

- high/low temperature

- dust, fume or mist

- radiation – ionising and non-ionising.

Examples of equipment with machinery hazards:

- Photocopier
- Document shredder
- Bench-top grinding machine
- Pedestal drill
- Cylinder mower
- Strimmer/Brush-cutter
- Chainsaw
- Compactor
- Checkout conveyor system
- Cement mixer
- Bench-mounted circular saw.

CONTROL MEASURES FOR REDUCING RISKS FROM MECHANICAL HAZARDS

Practical safeguarding

Defined by the legal definition of 'practical' – if it is technically possible to guard an item of equipment then it should be guarded irrespective of cost and convenience.

Levels of protection or hierarchy of measures

(Regulation 11 of PUWER)

1. Fixed enclosing guarding
2. Other guards or protection devices (interlocked, trip etc.)
3. Protection appliances (jigs, push-sticks etc.)
4. Provision of information, instruction, training and supervision.

TYPES OF GUARD

Guard	Example
Fixed	around a fan
Adjustable	around a drill bit on a pedestal drill
Interlocking	on the gates of a passenger lift, microwave oven
Trip device	photoelectric cell in front of a guillotine
Two-handed control	crimping machine
Hold-to-run	Tube train driver

Apply the above to the 11 pieces of equipment listed above.

FACTORS IN GUARD CONSTRUCTION

- Strength
- Weight and size
- Compatibility with process
- Hygiene
- Visibility
- Noise attenuation
- Free flow of air (ventilation and/or cooling)
- Free of sharp edges
- Maintenance and cleaning
- Openings.

Electrical Safety

2.5

LEARNING OUTCOMES

Outline the principles, hazards and risks associated with the use of electricity in the workplace

Outline the control measures that should be taken when working with electrical systems or using electrical equipment in normal workplace conditions

🔑 KEY REVISION POINTS

The basic electrical principles and definitions ☐

The causes and treatment of electric shock ☐

Other electrical hazards and injuries and their associated controls ☐

The selection, suitability, maintenance and inspection of electrical equipment ☐

The advantages and limitations of protective systems ☐

The hazards and controls associated with buried and overhead electrical services ☐

The application, advantages and limitations of portable electrical appliance testing ☐

HAZARDS AND RISKS ASSOCIATED WITH THE USE OF ELECTRICITY AT WORK

ELECTRICAL PRINCIPLES AND DEFINITIONS

■ voltage (volts V)

■ current (amps A)

■ resistance (ohms Ω)

■ Ohm's law – $V = I \times R$

■ Power $P = V \times I$ Watts

■ conductors – electricity will flow easily

■ insulators – very poor conductors

■ earthing – connection to earth

■ short circuit – direct flow to earth

■ low voltage – not exceeding 600 V

■ high voltage – exceeding 600 V

■ mains voltage – 220/240 V

ELECTRICAL HAZARDS AND INJURIES

■ electric shock

■ electric burns

■ electric fires and explosions

■ arcing

■ portable electrical equipment

■ secondary hazards (falls, trips, noise and vibration).

ELECTRIC SHOCK

■ The effect of an electric shock can vary from a slight tingling sensation to death;

■ The severity of the shock depends on the size of the current and voltage;

■ Low-voltage shocks, below 110 V, are seldom fatal;

■ Electric shock can also produce burns.

Treatment of electric shock:

■ Raise the alarm

■ Switch off the power – if not possible, use insulating material to move victim from contact with the power supply

■ Call for an ambulance

■ If breathing, place victim in recovery position

■ If not breathing, apply mouth-to-mouth resuscitation

■ Treat any burns

■ If person regains consciousness, treat for normal shock

■ Remain with person until help arrives.

If high voltages are involved, inform police and electrical supply company but do not approach the victim.

ELECTRICAL FIRES AND EXPLOSIONS

Twenty-five per cent of all fires have an electrical origin.

Causes:

- short circuits
- overheating of cables and equipment
- ignition of flammable gases and vapours
- ignition of combustible substances by static electrical discharges
- electric arcing
- static electricity – lightning strikes.

PORTABLE ELECTRICAL EQUIPMENT

Twenty-five per cent of all electrical accidents are caused by portable electrical equipment.

Hazards:

- faulty cables, extension leads, plugs and sockets
- inadequate maintenance
- use in flammable or damp atmospheres
- misuse of equipment.

SOURCES OF HIGH RISKS ASSOCIATED WITH THE USE OF ELECTRICITY

- working with poorly maintained electrical equipment
- using electrical equipment in adverse or hazardous environments such as wet, flammable or explosive atmospheres
- working on mains electricity supplies
- contact with underground cables during excavation work
- contact with live overhead power lines.

CONTROL MEASURES

The main control measures are given in the Electricity at Work Regulations, which cover the following topics:

- the design, construction and maintenance of electrical systems, work activities and protective equipment;

- the strength and capability of electrical equipment;

- the protection of equipment against adverse and hazardous environments;

- the insulation, protection and placing of electrical conductors;

- the earthing of conductors and other suitable precautions;

- the integrity of referenced conductors;

- the suitability of joints and connections used in electrical systems;

- means for protection from excess current;

- means for cutting off the supply and for isolation;

- the precautions to be taken for work on equipment made dead;

- working on or near live conductors;

- adequate working space, access and lighting; and

- the competence requirements for persons working on electrical equipment to prevent danger and injury.

SPECIFIC CONTROLS FOR ELECTRICAL HAZARDS

- A management system that ensures safe installation, operation and maintenance

- Training at all levels – induction, supervisory and technical

- Safe systems of work include permits to work on live electricity

- Selection of suitable equipment for the work and environment by considering the following:

    - atmosphere (flammable, damp)

    - weather conditions

    - high or low temperatures

    - dirty or corrosive conditions

    - standard of installation

    - rated operating conditions

- The use of protective systems

- Inspection and maintenance strategies.

Protective systems:

- Fuse (and circuit breaker)

- Insulation

- Isolation

- Reduced low voltage systems

- Residual current devices

- Double insulation

PROTECTION AGAINST CONTACT WITH LIVE OVERHEAD POWER LINES

As a general rule no vehicles, plant or equipment should be brought closer than:

- 15 m of overhead lines suspended from steel towers

- 9 m of overhead lines supported on wooden poles

- if a closer approach is required, lines should be made dead and a permit-to-work used.

If vehicles need to move beneath power lines:

■ the roadway should be covered by goalposts covered with warning tape;

■ bunting should be suspended, level with the top of the goalposts and just above ground level between poles across the site along the length of the power line;

■ suitable warning signs should be placed on either side of the roadway on each side of the power line to warn of the overhead power line; and

■ reach of vehicles/equipment should be restricted.

PROTECTION AGAINST CONTACT WITH LIVE BURIED POWER LINES

■ Check for any obvious signs of underground services, for example valve covers or patching of the road surface.

■ Ensure that the excavation supervisor has the necessary service plans and is competent to use them to locate underground services.

■ Ensure that all excavation workers are trained in safe digging practices and emergency procedures.

■ Use locators to trace any services and mark the ground accordingly. A series of trial holes should be dug by hand to confirm the position of the pipes or cables. This is particularly important in the case of plastic pipes, which cannot be detected by normal locating equipment.

■ In areas where underground services may be present, only hand digging should be used with insulated tools. Spades and shovels should be used rather than picks and forks, which are more likely to pierce cables.

■ Assume that all cables are 'live' unless it is known otherwise.

■ Hand-held power tools should not be used within 0.5 m of the marked position of an electricity cable. Collars should be fitted to the tools so that initial penetration of the surface is restricted.

- Any suspected damage to cables must be reported to the service providers and the health and safety enforcement authority.

- All exposed cables should be backfilled with fine material such as dry sand or small gravel.

- The service plans must be updated when the new cables have been laid.

INSPECTION AND MAINTENANCE STRATEGIES

The particular areas of interest for inspection and maintenance are:

- the competence of the inspection and maintenance team

- the cleanliness of insulator and conductor surfaces

- the mechanical and electrical integrity of all joints and connections

- the integrity of mechanical mechanisms, such as switches and relays

- the calibration, condition and operation of all protection equipment, such as circuit breakers, RCDs and switches

- isolation procedures

- inspection and maintenance records

- permits to work for work on live electricity.

INSPECTION STRATEGIES

Any strategy for the inspection of electrical equipment, particularly portable appliances, should involve the following considerations:

- a means of identifying the equipment to be tested

- the number and type of appliances to be tested

- the competence of those who will undertake the testing (whether in-house or bought-in)

- the legal requirements for portable appliance testing (PAT) and other electrical equipment testing and the guidance available

■ organisational duties of those with responsibilities for PAT and other electrical equipment testing

■ test equipment selection and recalibration

■ the development of a recording, monitoring and review system

■ the development of any training requirements resulting from the test programme.

PORTABLE ELECTRICAL APPLIANCE TESTING

General points:

■ user checks – at least once a week

■ formal visual inspection – by a trained person

■ combined testing and inspection – by a competent person using testing equipment

■ the frequency of inspection and testing depends on usage rate and conditions – normally determined by a risk assessment

■ records of inspection and testing required.

Advantages of portable appliance testing	Limitations of portable appliance testing
• early indication of serious faults	• too frequent testing of fixed equipment can lead to excessive costs
• discovery of incorrect equipment or electrical supply	• some unauthorised equipment never tested because of lack of records
• incorrect fuses found	• equipment misused or overused between tests
• reduction in the number of electrical accidents	• the inclusion of trivial faults leads to long lists and the overlooking of significant faults
• monitoring misuse of portable appliances	• the level of competence of tester
• check on selection procedures	• the use of incorrectly calibrated test equipment
• increase awareness of hazards	
• introduction of a more regular maintenance regime	

Fire Safety

LEARNING OUTCOMES

Describe the principles of fire initiation, classification and spread

Outline the principles of fire risk assessment

Describe the basic principles of fire prevention and the prevention of fire spread in buildings

Identify the appropriate fire alarm systems and fire-fighting equipment for a simple workplace

Outline the factors which should be considered when implementing a successful evacuation of a workplace in the event of a fire

KEY REVISION POINTS

The fire triangle ☐

The legal framework for the control of fire hazards ☐

The basic principles of fire prevention – classification, methods of extinction and principles of heat transmission ☐

The common causes of fire, its spread and consequences ☐

Key elements of a fire risk assessment ☐

The principles of fire detection, protection, warning and means of escape ☐

The provision, maintenance and testing of fire-fighting equipment ☐

Emergency planning and training ☐

FIRE INITIATION, CLASSIFICATION AND SPREAD

BASIC PRINCIPLES

The fire triangle consists of:

■ Fuel – solids, liquids and gases

■ Ignition source

■ Oxygen – from air, oxidising materials.

CLASSES OF FIRE

■ A – carboniferous solids

■ B1 – liquids soluble in water

■ B2 – liquids insoluble in water

■ C – gases

- D – metals
- F – high temperature cooking oils or fats.

Electrical fires do not constitute a separate fire class.

SOURCES OF IGNITION

- Naked flames
- External sparks
- Internal sparking
- Hot surfaces
- Static electricity.

METHODS OF EXTINCTION

- Cooling – reducing ignition temperature
- Smothering – limiting oxygen available
- Starving – limiting fuel supply
- Chemical reaction – interrupting the chain of combustion.

PRINCIPLES OF HEAT TRANSMISSION AND FIRE SPREAD

- Convection – through air movements (floor to floor)
- Conduction – through conducting materials (girders)
- Radiation – emission of heat waves (bar heaters)
- Direct burning – contact with flames (rubbish catching fire).

CAUSES OF FIRE

- rubbish (particularly if burnt)
- electrical short circuits

- portable heaters
- friction causing sparks (hand grinders)
- oil heating plant
- cooking appliances
- arson (often by children)
- hot surfaces (welding)
- cigarette smoking.

Fires spreads via:

- gases/sprays
- foam plastics
- packaging
- previous spillage
- flammable dusts.

Fire spreads due to:

- delayed discovery
- alarm not raised
- fire service not informed
- lack of barriers
- fire procedures not known
- fire extinguishers unavailable, not serviced or not used
- air flow – particularly if fire doors not closed
- lack of or poor emergency lighting
- flammable materials
- flammable dusts.

FIRE RISK ASSESSMENT

Required by the Regulatory Reform (Fire Safety) Order 2005.

STAGES OF THE FIRE RISK ASSESSMENT

1 Identify the fire hazards – identify sources of heat/ignition, unsafe acts (smoking) and unsafe conditions (storage of combustible materials);
2 Identify all locations and persons at risk;
3 Reduce the risks by the introduction of controls;
4 Record the findings; and
5 Monitor and review.

Hazards:

- sources of fuel

- oxygen depletion

- flames and heat

- smoke

- gaseous combustion products

- structural failure of the building.

Other factors:

- time to detect and warn

- safe egress

- adequacy of fire signs

- provision of fire-fighting equipment

- fire training – fire drills, appliance use and emergency teams

- maintenance of fire equipment

- consultation with employees.

TEMPORARY WORKPLACES

Construction sites, temporary classrooms and buildings, exhibitions, festivals and fetes.

The requirements depend on:

- the size of the premises;

- the number of persons working or visiting the site at any one time;

- the degree of flammability of the materials used to construct the workplace or being used in the workplace;

- height above or below the ground floor and distance to a place of safety; and

- the remoteness of the workplace relative to water supplies and/or fire and rescue services.

FIRE PREVENTION AND PREVENTION OF FIRE SPREAD

CONTROL MEASURES

- elimination or reduction in the use and storage of flammable materials

- electrical safety

- control of ignition sources

- effective systems of work

- safe refuelling procedures for vehicles

- safe heating and cooking equipment and procedures

- good housekeeping

- storage facilities for small quantities (50 litres) of highly flammable or flammable liquids.

FIRE ALARM AND FIRE-FIGHTING EQUIPMENT

FIRE EXTINGUISHERS

Fire extinguishers are coloured red over 95% of surface.

Colour (5%)	Contents	Class/type of fire
Red	Water	A
Cream	AFFF	A (possibly B)
Cream	Foam	B and C
Black	Carbon dioxide	B and electrical
Blue	Dry powder	B, C and electrical

OTHER FIRE-FIGHTING EQUIPMENT

- fire blankets
- hose reels
- sprinklers.

EMERGENCY LIGHTING

- Regular testing required.

SITING, MAINTENANCE AND TESTING OF FIRE FIGHTING EQUIPMENT

- sited in areas of significant risk (at least one extinguisher per 200 m^2 of floor space with a minimum of one per floor)
- arrangements for hot working
- annual maintenance check of fire extinguishers
- weekly check of fire alarms and emergency lighting.

Fire training in the use of the equipment for all staff is essential.

FIRE DETECTION AND WARNING

- fire notices
- alarms – including smoke and heat detectors
- audibility throughout premises
- regularly tested
- records of all tests
- vigilance of workers following training.

EVACUATION OF THE WORKPLACE

MEANS OF ESCAPE

- Doors – clear of obstructions and opening in direction of escape
- Escape routes (stairways and passageways) directly to a place of safety
- Lifts should not be used
- Lighting, including emergency lighting
- Directional and other signs
- Escape times
- Assembly points
- Building plans to include record of emergency escape plan.

STRUCTURAL DESIGN MEASURES

- Fire loading of the building
- Surface spread of fire
- Fire resistance of structural elements
- Insulating materials
- Fire compartmentation.

EMERGENCY PLANNING AND TRAINING

■ Evacuation procedures – fire notices

■ Appointment of fire marshals

■ Fire drills

■ Roll call

■ Provision for the infirm and disabled.

FIRE PROCEDURES FOR PEOPLE WITH A DISABILITY

Emergency planning needs to consider the special requirements of people with physical or mental disabilities. Such issues include:

■ the identification of those who may need special help to evacuate the building;

■ the allocation of responsibility to specific staff to identify and assist people with a disability during emergencies;

■ plan suitable escape routes;

■ have a system in place that enables a person with a disability to summon help in emergencies; and

■ include the arrangements for disabled persons in all staff fire training programmes.

GENERAL POINTS

■ Emergency procedures required for fire, explosions, flooding or structural collapse.

■ A hot work permit should include the need for a fire extinguisher and checks for signs of fire on completion of the work.

■ A hot work permit would be needed when cutting steel with an angle grinder and soldering pipework in a central heating system.

- Frost around the valve on a Liquid Petroleum Gas cylinder indicates that the valve is leaking.

- Cylinders containing Liquid Petroleum Gas should be stored in a locked external compound at least 3 m from any oxygen cylinders.

- The minimum distance that oxygen cylinders should be stored away from propane, butane and other inflammable gases is 3 m.

- Butane and propane are heavier than air and therefore represent a ground level fire hazard.

Chemical and Biological Health Hazards and Risk Control

LEARNING OUTCOMES

Outline the forms of, the classification of, and the health risks from exposure to hazardous substances

Explain the factors to be considered when undertaking an assessment of the health risks from substances commonly encountered in the workplace

Explain the use and limitations of Workplace Exposure Limits, including the purpose of long-term and short-term exposure limits

Outline control measures that should be used to reduce the risk of ill health from exposure to hazardous substances

Outline the hazards, risks and controls associated with specific agents

Outline the basic requirements related to the safe handling and storage of waste

🔑 KEY REVISION POINTS

The forms and classification of hazardous chemical and biological agents ☐

The functions of various human body organs and their vulnerability to hazardous substances ☐

The difference between acute and chronic effects ☐

The requirements of the COSHH Regulations – assessment and recommended control measures ☐

The definitions and limitations of workplace exposure limits (WELs) ☐

The principles of prevention and good practice ☐

The hierarchy of controls for hazardous substances ☐

The use of engineering controls (LEV) and personal protective equipment ☐

The health risks and controls associated with asbestos and other specific agents ☐

Health surveillance ☐

Emergency and maintenance controls ☐

The safe handling and storage of waste ☐

FORMS AND CLASSIFICATION OF, AND HEALTH RISKS FROM, HAZARDOUS SUBSTANCES

AGENTS

Chemical agents	Biological agents
• Dusts – including respirable dust (fine dust that remains in the lungs) • Gases • Vapours – substances very close to their boiling points (e.g. solvents) • Liquids • Mists – similar to vapours but closer to the liquid phase (e.g. paint sprays) • Fume – condensed metallic particles (e.g. welding fume)	• Fungi • Moulds • Bacteria • Viruses

CLASSIFICATION OF HAZARDOUS SUBSTANCES

▣ Irritant – repeated contact with skin can cause a sensitised or allergic inflammation

▣ Harmful – involves limited health risks

▣ Corrosive – may destroy living tissue

▣ Toxic – produces serious health risks

▣ Infectious – contains micro-organisms known to cause disease

▣ Carcinogenic – may induce cancer

▣ Mutagenic – may induce hereditary genetic defects.

Acute effects are of short duration, normally reversible and appear shortly after exposure to a hazardous substance (e.g. asthma attacks).

Chronic effects develop over a period of time after repeated exposure to a hazardous substance and are often irreversible.

ASSESSMENT OF HEALTH RISKS

Categories of health risks	Types of health risk
• chemical – exhaust fumes, paint solvents, asbestos • biological – Legionella, other pathogens, hepatitis	• skin contact with irritant substances, leading to dermatitis; • inhalation of respiratory sensitisers, triggering immune responses such as asthma;
• physical – noise vibration, radiation • psychological – stress, violence • (ergonomic – musculoskeletal disorders)	• badly designed workstations requiring awkward body postures or repetitive movements, resulting in upper limb disorders, repetitive strain injury and other musculoskeletal conditions; • noise levels which are too high, causing deafness and conditions such as tinnitus; • too much vibration, for example from hand-held tools leading to hand–arm vibration syndrome and circulatory problems; • exposure to ionising and non-ionising radiation including ultraviolet in the sun's rays, causing burns, sickness and skin cancer; • infections ranging from minor sickness to life-threatening conditions, caused by inhaling or being contaminated with microbiological organisms; and • stress causing mental and physical disorders.

ROUTES OF ENTRY INTO THE BODY

■ Inhalation via the lungs

■ Absorption via the skin

■ Ingestion via the stomach

■ Injection through the skin.

MAJOR HUMAN BODY SYSTEMS

There are five major functional systems within the body. These are shown together with typical diseases and related hazardous substances:

Body system	Illness or disease	Hazardous substance
Respiratory	bronchitis, asthma, fibrosis, cancer	dusts, asbestos
Nervous	anxiety, epilepsy, loss of consciousness	solvents, lead
Cardiovascular	headaches, loss of consciousness	benzene, carbon monoxide
Urinary	cirrhosis, cancer	chlorinated hydrocarbons, mercury
Skin	irritant contact dermatitis and allergic contact dermatitis	detergents, turpentine, epoxy resin

CONTROL OF SUBSTANCES HAZARDOUS TO HEALTH REGULATIONS

REQUIREMENTS OF COSHH

■ Suitable and sufficient health risk assessment

■ Adequate control of exposure of employees

■ Proper use of control measures provided

■ Maintain control measures

■ Monitor employees exposed to hazardous substances (Schedule 5)

■ Health surveillance (Schedule 6)

■ Information, instruction and training.

COSHH ASSESSMENT

1 Identify hazardous substances;
2 Gather information;
3 Evaluate the risks;
4 Decide on control measures;
5 Record the assessment; and
6 Review the assessment.

SOURCES OF INFORMATION

■ Product labels

■ Safety data sheets

■ HSE Guidance Note EH 40

■ Trade association publications

■ Industrial codes of practice

■ Specialist reference manuals.

SURVEY TECHNIQUES FOR HEALTH RISKS

include:

■ Stain tube detector. The advantages of this technique are that it is quick, relatively simple to use and inexpensive. The disadvantages are:

- it cannot be used to measure concentrations of dust or fume

- the accuracy of the reading is approximately $\pm$ 25%

- it will yield false readings if other contaminants present react with the crystals

- it can only give an instantaneous reading, not an average reading over the working period (TWA)

- the tubes are very fragile with a limited shelf life

■ Passive sampling, including the static dust sampler

■ Sampling pumps and heads

■ Direct reading instruments (MIRA)

■ Vane anemometers and hygrometers

■ Smoke tube and dust observation lamp.

WORKPLACE EXPOSURE LIMITS (WELs)

Workplace exposure limits (WELs) must not be exceeded.

Two categories (similar to previous MEL and OES):

Category 1	carcinogenic and mutagenic substances – exposure must be reduced as low as is reasonably practicable below the WEL
Category 2	all other hazardous substances – exposure controlled by principles of good practice

There are two time-weighted averages (TWA):

■ the long-term exposure limit (LTEL) – over 8 hours

■ the short-term exposure limit (STEL) – 15-minute reference period.

LIMITATIONS OF EXPOSURE LIMITS

■ They are specifically quoted for an 8-hour period (with an additional STEL for many hazardous substances). Adjustments must be made when exposure occurs over a continuous period longer than 8 hours.

■ They can only be used for exposure in a workplace and not to evaluate or control non-occupational exposure (e.g. to evaluate exposure levels in a neighbourhood close to the workplace, such as a playground).

■ WELs are only approved where the atmospheric pressure varies from 900 to 1100 millibars. This could exclude their use in mining and tunnelling operations.

■ They should also not be used when there is a rapid build-up of a hazardous substance due to a serious accident or other emergency.

■ The fact that a substance has not been allocated a WEL does not mean that it is safe.

CONTROL MEASURES

The general principles of prevention are:

1 the avoidance of risks;
2 the evaluation of unavoidable risks;
3 controlling hazards at source; and
4 replacing the dangerous by less or non-dangerous alternatives;
5 adapting work to the individual;
6 adapting to technical progress;
7 developing a coherent prevention policy;
8 giving priority to collective over individual protective measures; and
9 giving appropriate instructions to employees.

PRINCIPLES OF GOOD PRACTICE FOR THE CONTROL OF EXPOSURE TO SUBSTANCES HAZARDOUS TO HEALTH

1 Design and operate processes and activities to minimise emission, release and spread of substances hazardous to health.
2 Take into account all relevant routes of exposure – inhalation, skin absorption and ingestion – when developing control measures.
3 Control exposure by measures that are proportionate to the health risk.
4 Choose the most effective and reliable control options that minimise the escape and spread of substances hazardous to health.
5 Where adequate control of exposure cannot be achieved by other means, provide, in combination with other control measures, suitable PPE.
6 Check and review regularly all elements of control measures for their continuing effectiveness.
7 Inform and train all employees on the hazards and risks of the substances with which they work and the use of control measures developed to minimise the risks.

8 Ensure that the introduction of control measures does not increase the overall risk to health and safety.

HIERARCHY OF CONTROL MEASURES

1 eliminate the risks by either designing them out or changing the process;
2 substitution of hazardous substances with others which are less hazardous;
3 isolation, using enclosures, barriers or worker segregation;
4 engineering controls such as guarding, the provision of local exhaust ventilation systems, the use of reduced voltage systems or residual current devices;
5 management controls such as safe systems of work, training, job rotation and supervision; and
6 finally, as a last resort, the provision of personal protective equipment such as ear defenders or respiratory protective equipment.

Hence, measures for preventing or controlling exposure to hazardous substances include one or a combination of the following:

■ elimination of the substance

■ substitution of the substance (or reduction in the quantity used)

■ total or partial enclosure of the process

■ local exhaust ventilation

■ dilute or general ventilation

■ reduction of the number of employees exposed to a strict minimum

■ reduced time exposure by task rotation and the provision of adequate breaks

■ good housekeeping

■ training and information on the risks involved

■ effective supervision to ensure that the control measures are being followed

■ personal protective equipment (such as clothing, gloves and masks)

■ welfare (including first aid)

■ medical records

■ health surveillance.

PREVENTATIVE CONTROL MEASURES

ENGINEERING CONTROLS

These include:

- segregation of people from the process (fume cupboard)

- modification of the process to reduce human contact with the hazardous substances

- local exhaust ventilation comprising:

  - a collection hood and intake

  - ventilation ducting

  - a filter or other air cleaning device

  - a fan

  - an exhaust duct

- dilution or general ventilation.

SUPERVISORY OR PEOPLE CONTROLS

Additional supervisory controls when hazardous substances are involved are:

- reduced time exposure – thus ensuring that workers have breaks in their exposure periods;

- reduced number of workers exposed – only persons essential to the process should be allowed in the vicinity of the hazardous substance. Walkways and other traffic routes should avoid any area where hazardous substances are in use;

- eating, drinking and smoking must be prohibited in areas where hazardous substances are in use;

- any special rules, such as the use of personal protective equipment, must be strictly enforced.

PERSONAL PROTECTIVE EQUIPMENT

Personal protective equipment (PPE) is the control measure of **last resort**. It is covered by the Personal Protective Equipment at Work Regulations, which require:

■ PPE to be suitable for wearer and task;

■ PPE to be compatible with any other personal protective equipment;

■ a risk assessment to be undertaken to determine best PPE;

■ a suitable PPE maintenance programme;

■ suitable storage arrangements for the PPE;

■ information, instruction and training for the user of the PPE;

■ supervision of the use of PPE; and

■ a system for reporting defects.

Notes:

1. The Regulations do not cover Respiratory Protective Equipment (RPE).
2. The risk assessment and method statement for the particular activity should include information on personal protective equipment.

Types of personal protective equipment	Types of respiratory protective equipment
• hand and skin protection • eye protection • protective clothing – boots, hard hats, aprons	• filtering half mask • half mask respirator • full face mask respirator • powered respirator

HEALTH SURVEILLANCE

Health surveillance is not the same as health monitoring. Types of health surveillance include:

■ checking individuals on a regular basis;

■ checking individuals as a result of statistical evidence of a potential problem; and

■ when a substance listed in Schedule 6 of COSHH is being used. In these cases, health surveillance should take place at intervals not exceeding 12 months and records should be kept for 40 years.

MAINTENANCE AND EMERGENCY CONTROLS

Maintenance involves:

■ the cleaning, testing and, possibly, dismantling of equipment;

■ the changing of filters in extraction plant or entering confined spaces;

■ the handling of hazardous substances and the safe disposal of waste material;

■ a permit-to-work procedure may also need to be in place, since the control equipment will be inoperative during the maintenance operations;

■ the keeping of records of maintenance for at least 5 years.

The following points should be considered when **emergency procedures** are being developed:

■ the possible results of a loss of control (e.g. lack of ventilation);

■ dealing with spillages and leakages (availability of effective absorbent materials);

■ raising the alarm for more serious emergencies;

■ evacuation procedures, including the alerting of neighbours;

■ fire-fighting procedures and organisation;

■ availability of respiratory protection equipment; and

■ information and training.

SPECIFIC AGENTS

ASBESTOS

A fine, sharp and hard fibrous dust of respirable dust size which can become lodged in the lungs, causing damage to the lining of the lungs over a period of many years. Asbestos is a possible chronic health risk. It can lead to one of the following diseases:

■ asbestosis or fibrosis (scarring) of the lungs

■ lung cancer

■ mesothelioma – cancer of the lining of the lung.

Duty to manage asbestos

The Control of Asbestos at Work Regulations require:

■ the determination of the location and condition of materials likely to contain asbestos (ACMs)

■ the presumption that materials contain asbestos unless there is strong counter-evidence

■ up-to-date records of the location and condition of the ACMs or presumed ACMs

■ an assessment of the risk of anyone being exposed to fibres from ACMs

■ a plan to manage the risks from ACMs

■ the necessary steps to action the plan

■ the periodic review and monitoring of the plan

■ provision of information and asbestos awareness training to anyone who is liable to work on ACMs or otherwise disturb them.

Types of asbestos survey

■ the presumptive survey – the location and assessment of condition of ACMs

■ the sampling survey – a number of samples are sent away for analysis

■ the full access and sampling survey – most invasive and detailed.

Stages in the control of asbestos

■ **Identification** of the presence of asbestos

■ **Assessment** is an evaluation as to whether the location or the condition could lead to the asbestos being disturbed

■ **Removal** must only be done by a licensed contractor

■ **Control measures** during the removal of asbestos to ensure adequate protection of workers and others

■ **Medical surveillance** in the form of a regular medical examination should be given to any employee who has been exposed to asbestos at levels above the action level

■ **Awareness training** is given to those employees:

  ● who are or who are liable to be exposed to asbestos or who supervise such employees

  ● who carry out work in connection with the employer's duties under these Regulations, so that they can carry out that work effectively

■ **Disposal of asbestos waste**: to an authorised asbestos waste site.

If asbestos is discovered during the performance of a contract, work should cease immediately and the employer should be informed.

HEALTH RISKS AND CONTROLS ASSOCIATED WITH OTHER SPECIFIC AGENTS

Each agent is followed by a brief description, two possible illnesses and possible controls.

Hepatitis and many blood–borne viruses can cause nausea, jaundice and death. Possible control measures:

■ prohibit eating, drinking and smoking in working areas where there is a risk of contamination;

■ prevent puncture wounds, cuts and abrasions, especially in the presence of blood and body fluids;

- when possible avoid use of, or exposure to, sharps such as needles, glass or metal, or if unavoidable take care in handling and disposal;

- consider the use of devices incorporating safety features, such as safer needle devices and blunt-ended scissors;

- cover all breaks in exposed skin by using waterproof dressings and suitable gloves;

- protect the eyes and mouth by using a visor/goggles/safety spectacles and a mask where splashing is possible;

- avoid contamination by using water-resistant protective clothing;

- wear rubber boots or plastic disposable overshoes when the floor or ground is likely to be contaminated;

- use good basic hygiene practices, such as hand washing;

- control contamination of surfaces by containment and using appropriate decontamination procedures; and

- dispose of contaminated waste safely.

HIV is an example of a blood-borne virus.

Tetanus – a bacterium in soil enters a wound from an infected object and attacks the nerves that supply muscle tissue (lockjaw); can be fatal.

Organic solvents are sensitisers and irritants that can cause dermatitis and liver failure. The minimum PPE requirements are impermeable overalls, apron, footwear, long gloves and gauntlet and chemically resistant goggles or visor. Respiratory protective equipment is also required if it cannot be demonstrated that exposure is below the appropriate workplace exposure limit (WEL). Organic solvents are often used in painting work.

Carbon monoxide is an odourless gas that can cause headaches and unconsciousness. Exposure may be prevented by ensuring that any work carried out in relation to gas appliances in domestic or commercial premises is undertaken by a competent Gas Safe Registered engineer.

Isocyanates are sensitisers and irritants that can cause bronchitis and extreme 'asthma attack'. They are often used in paint-spraying operations and require good ventilation. Respiratory Protective Equipment (RPE) should be worn.

Cement dust and wet cement can cause serious burns or ulcers and dermatitis. PPE in the form of gloves, overalls with long sleeves and full-length trousers and waterproof boots must be worn.

Wood dust can be hazardous, particularly from hard woods and composite boards (MDF), and can cause irritation to the eyes, nose and throat, leading to dermatitis, asthma, rhinitis and even cancer. An extraction system is essential and PPE in the form of gloves, respiratory protective equipment overalls and eye protection.

Silica dust produced in masonry work can cause fibrosis and silicosis. Prevention is best achieved by the use of good dust extraction systems and respiratory protective equipment.

Leptospira is a bacterium in rat urine that can cause anaemia and jaundice. Good, impervious protective clothing, particularly Wellington boots, and the covering of any skin wounds are essential in these situations.

Legionella are bacteria in tepid water that can cause pneumonia and death. Where a plant is at risk for the development of Legionella, the following is required:

- a written 'suitable and sufficient' risk assessment;

- the preparation and implementation of a written control scheme involving the treatment, cleaning and maintenance of the system;

- appointment of a named person with responsibility for the management of the control scheme;

- the monitoring of the system by a competent person; and

- record keeping and the review of procedures developed within the control scheme.

SAFE HANDLING AND STORAGE OF WASTE

Mainly regulated by:

■ The Hazardous Waste Regulations – Special Waste Regulations still in Scotland

■ The Environmental Protection Act (EPA)

The principal requirements of the EPA are as follows:

■ to handle waste so as to prevent any unauthorised escape into the environment;

■ to pass waste only to an authorised person;

■ to ensure that a written description accompanies all waste; and

■ to ensure that no person commits an offence under the Act.

There are several different categories of waste. The important ones are as follows:

■ Controlled waste, comprising household, industrial or commercial waste

■ Hazardous waste, which can only be disposed of using special arrangements (toxic, corrosive, carcinogenic or highly flammable).

REQUIREMENTS FOR WASTE SKIPS

■ sufficient strength to cope with its load;

■ stability while being filled;

■ a reasonable uniform load distribution within the skip at all times;

■ the immediate removal of any damaged skip from service and the skip inspected after repair before it is used again;

■ sufficient space around the skip to work safely at all times;

■ the skip should be resting on firm level ground;

- the skip should never be overloaded or overfilled; and

- there must be sufficient headroom for the safe removal of the skip when it is filled.

THE CONTROL MEASURES FOR HAZARDS ASSOCIATED WITH WASTE DISPOSAL

- All workers should wear suitable PPE, such as high-visibility jackets, gloves and suitable footwear.

- Training so that employees segregate hazardous and non-hazardous wastes on site and fully understand the risks and necessary safety precautions to be taken.

- The storage site should be protected against trespassers, fire and adverse weather conditions.

- If flammable or combustible wastes are being stored, adequate fire protection systems must be in place.

- Manual handling issues may need to be considered.

- All lifting equipment, including chains and shackles, must be subject to a periodic statutory examination.

- Separate storage required for incompatible waste streams.

- Finally, for liquid wastes, drains must be protected and bunds used to restrict spreading of the substance as a result of spills.

Physical and Psychological Health Hazards and Risk Control

2.8

LEARNING OUTCOMES

Outline the health effects associated with exposure to noise and appropriate control measures

Outline the health effects associated with exposure to vibration and appropriate control measures

Outline the health effects associated with ionising and non-ionising radiation and outline appropriate control measures

Outline the causes and effects of stress at work and appropriate control measures

PHYSICAL AND PSYCHOLOGICAL HEALTH HAZARDS AND RISK CONTROL

🔑 KEY REVISION POINTS

The health effects of exposure to noise and vibration ☐

The legal exposure limits to noise and vibration ☐

The control measures for excessive vibration and noise ☐

The hazards and controls associated with ionising and non-ionising radiations ☐

The causes and prevention of workplace stress ☐

NOISE

Human ear:

- Cochlea

- Hair cells – damage irreversible

- Auditory nerve.

EFFECTS OF NOISE

Acute effects of noise:

- Temporary threshold shift

- Tinnitus

- Acute acoustic trauma.

Chronic effects of noise:

- Noise-induced hearing loss

- Permanent threshold shift

- Tinnitus.

NOISE ASSESSMENTS

Objective is to identify whether action levels have been reached.

NOISE MEASUREMENT

■ Sound pressure level (SPL) – dB(A)

■ Peak sound pressure

■ Continuous equivalent noise level (L_{Eq})– normally measured over an 8-hour period

■ Daily personal exposure level ($L_{EP,d}$).

REQUIREMENTS OF NOISE AT WORK REGULATIONS

■ assess noise levels and keep records;

■ reduce the risks from noise exposure by engineering controls – only use hearing protection as a last resort;

■ provide employees with information and training; and

■ if a manufacturer of equipment, to provide information on noise levels of the equipment.

Noise action levels:

■ First action level – daily exposure level of 80 dB(A) – action advised

■ Second action level – daily exposure level of 85 dB(A) – action obligatory

■ Peak action level – 135 dB(C) and 137 dB(C)

■ Exposure limit values – 87 dB(A) and 140 dB(C).

Noise control techniques include:

■ Reduction of noise at source

■ Change equipment or process

■ Change speed

■ Improve maintenance

- Relocate equipment
- Enclose equipment
- Screens or absorption walls
- Damping
- Lagging
- Silencers
- Isolation of workers
- Suitable warning signs.

Personal ear protection:

- Ear plugs
- Ear defenders (earmuffs).

Factors to be considered:

- Suitability (frequencies)
- Acceptability and comfort
- Durability
- Instruction in use
- Hygiene considerations
- Beards, hair and spectacles may reduce effectiveness of protection
- Maintenance and storage
- Cost.

VIBRATION

ILLHEALTH DUE TO VIBRATION

Illhealth due to hand–arm vibration

- Hand–arm vibration syndrome (HAVS)

- Vibration White Finger (VWF)
- Carpal tunnel syndrome.

Ill-health due to whole body vibration

- Severe back pain
- Reduced visual and manual control
- Increased heart rate and blood pressure
- Spinal damage.

CONTROL OF VIBRATION AT WORK REGULATIONS

Exposure limit and action values

For hand–arm vibration (HAV):

- the daily exposure limit value normalised to an 8-hour reference period is 5 m/s^2
- the daily exposure action value normalised to an 8-hour reference period is 2.5 m/s^2.

For whole body vibration (WBV):

- the daily exposure limit value normalised to an 8-hour reference period is 1.15 m/s^2
- the daily exposure action value normalised to an 8-hour reference period is 0.5 m/s^2.

PREVENTATIVE AND PRECAUTIONARY MEASURES

The common measures used to control ergonomic ill-health effects are:

- identification of repetitive actions;
- the elimination of vibration by performing the job in a different way;
- undertake a risk assessment;

- ensure that the correct equipment (properly adjusted) is always used;

- introduce job rotation or frequent breaks so that workers have a reduced time exposure to the hazard;

- during the design of the job ensure that poor posture is avoided;

- issue employees with gloves and warm clothing;

- examine ill-health reports and absence records;

- introduce a programme of health surveillance;

- ensure that workers are given adequate information on the hazards and develop a suitable training programme;

- introduce a reporting system for employees to use so that concerns and any symptoms can be recorded and investigated;

- ensure that drill bits and other tools are kept sharp;

- ensure that a programme of preventative maintenance is introduced and include the regular inspection of items such as vibration isolation mountings; and

- keep up to date with advice from equipment manufacturers, trade associations and health and safety sources.

ADDITIONAL INFORMATION ON WHOLE BODY VIBRATION

The reasons for back pain in drivers include:

- poor posture while driving

- incorrect adjustment of the driver's seat

- difficulty in reaching all relevant controls due to poor design of the controls layout

- frequent manual handling of loads

- frequent climbing up and down from a high cab.

Actions for controlling the risks from WBV need to ensure that:

■ the driver's seat is correctly adjusted so that all controls can be reached easily and that the driver weight setting on the suspension seat, if available, is correctly adjusted. The seat should have a back rest with lumbar support;

■ anti-fatigue mats are used if the operator has to stand for long periods;

■ the speed of the vehicle is such that excessive jolting is avoided. Speeding is one of the main causes of excessive WBV;

■ all vehicle controls and attached equipment are operated smoothly;

■ only established site roadways are used;

■ only suitable vehicles and equipment are selected to undertake the work and cope with the ground conditions;

■ the site roadway system is regularly maintained;

■ all vehicles are regularly maintained, with particular attention being paid to tyre condition and pressures, vehicle suspension systems and the driver's seat;

■ work schedules are regularly reviewed so that long periods of exposure on a given day are avoided and drivers have regular breaks;

■ prolonged exposure to WBV is avoided for at risk groups (older people, young people, people with a history of back problems and pregnant women); and

■ employees are aware of the health risks from WBV, the results of the risk assessment and the ill-health reporting system. They should also be trained to drive in such a way that excessive vibration is reduced.

GENERAL POINTS ON VIBRATION

■ A person who is warm and dry is less likely to suffer from HAV.

■ HAV can cause damage to blood vessels and nerves in hands and fingers.

■ The effects of HAV can be reduced if the work is done in short intervals.

■ Anti-vibration gloves may not always protect against vibration.

RADIATION

RADIATION HAZARDS

Ionising radiation

Ionising radiation is produced by alpha and beta particles and gamma rays.

Harmful effects:

- Somatic – cell damage to the individual

- Genetic – cell damage to the children of the individual

- Acute – nausea, vomiting, skin burns and blistering, collapse and death

- Chronic – anaemia; leukaemia and other types of cancer.

Sources:

- Radon gas

- X-ray equipment

- Smoke detectors.

Non-ionising radiation

- Ultraviolet radiation – sun, arc welding

- Lasers – eye and skin burns, electricity etc.

- Infra-red radiation – fires, furnaces etc.

- Microwaves – cookers/ovens, mobile telephones.

Non-ionising radiation can harm the eyes and skin.

RADIATION PROTECTION STRATEGIES

Ionising radiation

Covered by the Ionising Radiations Regulations, which specify:

- Risk assessment

- Shielding

- Time (reduced time exposure)
- Distance
- Training
- Personal protective equipment
- No food or drink consumption near exposed areas
- Signs and information
- Medical surveillance
- Maintenance and inspection controls
- Emergency procedures
- Radiation protection supervisor – appointed to advise on compliance with Regulations
- Radiation protection adviser – appointed to advise the radiation protection supervisor.

Non-ionising radiation

- Eye protection
- Skin protection (gloves and/or creams)
- Fixed shields and non-reflective surfaces
- Interlocking guards on microwaves.

STRESS

The reaction of the body to excessive mental pressure that may lead to ill-health.

Causes:

- The job itself – unrealistic targets, boring, repetitive, insufficient training
- Individual responsibility – ill-defined roles, too much responsibility, too little control to influence outcome

- Working conditions – lack of privacy or security, unsafe practices, threat of violence, excessive noise

- Management attitudes – negative health and safety culture, poor communication, lack of support in a crisis

- Relationships with colleagues – bullying, harassment.

Prevention:

- Take a positive attitude to stress

- Take employee concerns seriously

- Effective communication and consultation

- Develop a policy on stress

- Relevant training

- Employee appraisal system

- Discourage excessive hours at work

- Encourage lifestyle changes

- Monitor incidents of bullying etc.

- Avoid blame culture

- Set up a confidential counselling advice service.

Useful Additional Construction Topics

THE SCOPE OF CONSTRUCTION

The scope of the construction industry includes:

- general building work – domestic, commercial or industrial, and may be:

  - new building work, such as a building extension, or

  - the refurbishment, maintenance or repair of existing buildings;

- civil engineering projects involving road and bridge building, water supply and sewage schemes, and river and canal work;

- the use of woodworking workshops together with woodworking machines and their associated hazards;

- electrical installation and plumbing work;

- painting and decorating; and

- work in confined spaces such as excavations and underground chambers.

CONSTRUCTION HAZARDS AND CONTROLS

1. SAFE PLACE OF WORK

- secure and locked gates with appropriate notices posted;

- a secure and undamaged perimeter fence with appropriate notices posted;

- all ladders either stored securely or boarded across their rungs;

- all excavations covered;

■ all mobile plant immobilised, where practicable, and services isolated;

■ secure storage of all inflammable and hazardous substances;

■ visits to local schools to explain the dangers present on a construction site;

■ if unauthorised entry persists, then security patrols and closed circuit television may need to be considered.

2. PROTECTION AGAINST FALLS

The Work at Height Regulations give the following hierarchy of control:

1　Avoid working at height if possible
2　Use an existing safe place of work
3　Provide work equipment to prevent falls
4　Mitigate distance and consequences of a fall
5　Instruction, training and supervision.

Following a suitable risk assessment, the following hierarchy of measures should be considered:

■ avoid working at height, If possible;

■ the provision of a properly constructed working platform, complete with toe boards and guardrails;

■ if this is not practicable or where the work is of short duration, suspension equipment should be used, and only when this is impracticable;

■ collective fall arrest equipment (air bags or safety nets) may be used;

■ where this is not practicable, individual fall restrainers (safety harnesses) should be used;

■ only when none of the above measures are practicable should ladders or stepladders be considered.

3. FRAGILE ROOFS

Particular hazards are:

- fragile roofing materials – more brittle with age and exposure to sunlight;

- exposed edges;

- unsafe access equipment;

- falls from girders, ridges or purlins;

- overhead services and obstructions;

- the use of equipment such as gas cylinders and bitumen boilers; and

- manual handling hazards.

Controls include:

- suitable means of access such as scaffolding, ladders and crawling boards;

- suitable barriers, guard rails or covers where people work near to fragile materials and roof lights; and

- suitable warning signs indicating that a roof is fragile, at ground level.

4. PROTECTION FROM FALLING OBJECTS

- the use of covered walkways or suitable netting to catch falling debris;

- waste material should be brought to ground level by the use of chutes or hoists;

- minimal quantities of building materials should be stored on working platforms; and

- hard hats must be given to all employees whenever there is a risk of head injury from falling objects. Self-employed workers must supply their own head protection. Visitors to construction sites should be supplied with head protection and mandatory head protection signs displayed around the site.

5. DEMOLITION CONTROLS

Include:

- prior to work, a full site investigation;

- risk assessments (two – project designer and contractor);

- a method statement;

- waste disposal arrangements;

- piecemeal and deliberate controlled collapse; and

- training and information to all workers.

6. PREVENTION OF DROWNING

Arrangements in place to prevent people falling into the water and ensure that rescue equipment is available at all times.

7. VEHICLES AND TRAFFIC ROUTES

- all vehicles must be well maintained and only driven by trained persons;

- traffic routes, loading and storage areas need to be well designed with enforced speed limits, good visibility and the separation of vehicles and pedestrians being considered;

- the use of one-way systems and separate site access gates for vehicles and pedestrians may be required; and

- the safety of members of the public must be considered, particularly where vehicles cross public footpaths.

8. FIRE AND OTHER EMERGENCIES

Emergency procedures are required for fire, explosions, flooding or structural collapse and should include:

- the location of fire points and assembly points;

- accident reporting and investigation; and

- rescue from excavations and confined spaces.

9. WELFARE FACILITIES

To include:

- sanitary and washing facilities (including showers if necessary) with an adequate supply of drinking water;

- accommodation for the changing and storage of clothes;

- rest facilities for break times;

- adequate first aid provision (an accident book); and

- protective clothing against adverse weather conditions.

10. ELECTRICITY

- only 110 V equipment should be used on site;

- if mains electricity is used, then residual current devices should be used with all equipment;

- where workers or tall vehicles are working near or under overhead power lines, either the power should be turned off or 'goal posts' or taped markers used to prevent contact with the lines; and

- similarly, underground supply lines should be located and marked before digging takes place.

11. NOISE

- noisy machinery should be fitted with silencers;

- ear defenders issued when working with noisy machinery; and

- when machinery is used in a workshop (such as woodworking machines), a noise survey should be undertaken.

12. ADDITIONAL HEALTH HAZARDS

Include:

- dust – including asbestos;

- vibration;

- cement dust and wet cement;

- wood dust;

- solvents, cleaners and paints;

- tetanus and other biological agents; and

- manual handling.

13. WASTE DISPOSAL

- the disposal of waste must be properly managed;

- waste skips must not be overloaded or used for inappropriate waste.

Specimen Answers to NEBOSH Examination Questions

INTRODUCTION

This chapter is only relevant to those readers who are due to sit for NEBOSH examinations.

The accompanying textbook to this Revision Guide – *Introduction to Health and Safety at Work* – gives detailed advice on studying for NEBOSH examinations and some sample solutions to NEBOSH examination questions. It also provides a sample practical assessment. To gain the full benefit from this section of the Revision Guide, those chapters of the textbook should also be read. But full success in the examinations is only likely to be achieved if a good examination technique is adopted.

FEATURES OF A GOOD EXAMINATION TECHNIQUE

A. MNEMONICS

The use of mnemonics during your revision period can be very helpful in the examination. A mnemonic can be a sentence or a word that enables a person to remember a list of items. For example, the colours of the rainbow – red, orange, yellow, green, blue, indigo and violet – can be remembered by the sentence 'Richard Of York Gained Battles In Vain'. Similarly, the main factors in a manual handling assessment can be remembered by the word TILE – Task, Individual, Load and Environment.

The best mnemonics are those which students have devised for themselves. The reader is advised to invent his or her own set of mnemonics based on familiar words or sentences. Some examples of useful mnemonics are given below.

1. Management of health and safety

Elements in HSG65	Personal factors	Common topics in the management paper
Policy	Self-interest	Safety culture
Organisation	Hearing/memory loss	Training
Plan	Experience	Risk assessment
Monitor	Age	Information
Audit	Training level	Maintenance
Review	Health	Supervision
		Safe system of work

POPMAR, SHEATH and **STRIMSS** are stand-alone words. (STRIMMS was devised by a student from the horticultural industry.)

2. Controlling workplace hazards

Risk assessment stages (inc. COSHH and noise)	Machine hazards	Machine guarding
Assess the hazards	Entanglement	Fixed or fixed distance
Control the risks	Nips	Interlocked
Monitor the controls	Traps	Adjustable
Inform employees	Impact	Trip devices
Record and review the assessment	Contact or cutting	
	Ejection	

ACMIR can be remembered by 'All Colours Must Include Red'. **ENTICE** and **FIAT** are stand-alone words.

It is very important to stress that a mnemonic is only an aid to memory, and not all the elements in a mnemonic may be relevant to a particular question. Where all or parts of a mnemonic are relevant to a question, you will need to expand on it in your answer.

For example, if the mnemonic, such as STRIMSS, reminds you that 'training and information' are relevant to the question, you should add a brief description of the training and information that you have in mind. Very few marks can be awarded for simply stating 'training and information'.

B. USE OF PERSONAL EXPERIENCE

The NEBOSH course covers much factual material which many students find difficult to remember. It is easier to remember this material if it can be related to personal experience and their own workplace. For example, most workplaces have some hazardous substances present, even if only as cleaning materials. These can be used to illustrate an answer involving hazardous substances (provided that the reference is relevant to the question!). Always try to use familiar examples of hazards and controls when answering examination questions involving those hazards and controls.

C. SIMPLE RULES WHEN TAKING THE EXAMINATION

1 Always arrive at least 10 minutes before the start of the examination.
2 Quickly read the whole paper before attempting any question.
3 Start each question on a separate sheet in the answer book and use both sides of the paper. The current NEBOSH answer book has the relevant question number printed on each page of the book.
4 Make brief notes at the beginning of the question page in the answer book.
5 Answer the easiest question first – this may not necessarily be the first one.
6 Read the question carefully and highlight the action verbs (e.g. outline, describe, state etc.). These action verbs are printed in **bold letters** in the question on the examination paper. It is common for candidates to answer the question that they would like to answer rather than the actual one on the examination paper.
7 Answer **all** parts of the question.
8 Attempt **all** questions – you can only get zero marks for an unanswered question.

9   Time is very limited during the examination and answers should be brief and to the point. Do not write all that you know about the topic being questioned or pad out your answer. You should spend no more than 25 minutes on the long answer question (covering about a page and a half) and 8 minutes on each short answer question (normally covering half a page). Most examination failures are caused by not allowing sufficient time to answer all the questions.

10 Finally, the answers should be easy for the examiner to read in terms of layout and standard of writing.

SPECIMEN ANSWERS TO NEBOSH EXAMINATION QUESTIONS

Some specimen answers are given below to past NGC1 and NGC2 examination questions. Detailed advice and an example of NGC3 – the Health and Safety Practical Application – are given in the textbook that accompanies this Revision Guide (*Introduction to Health and Safety at Work*).

It is recommended that, to gain full benefit from the specimen answers, the reader should attempt the questions initially without consulting the answers. The answers given are generally longer than would be expected in the examination because they have covered all possible options that are available in answering a particular question.

Finally, two long answer questions are provided for NGC1 because candidates often have more difficulty with this paper.

1. NGC1 – THE MANAGEMENT OF HEALTH AND SAFETY

Question 1

A workshop has experienced a number of accidents that were due to inadequate selection and implementation of risk controls.

(a) **Identify FOUR** of the general principles of prevention. **(4)**
(b) **Give** the meaning of the term 'hierarchy of control'. **(2)**
(c) **Outline** how a typical hierarchy of control should be applied. **(8)**

(d) **Outline** the factors to be considered when selecting an appropriate risk control method. **(6)**

Answer:

(a) Four general principles of prevention are:

1 the avoidance of risks;
2 the evaluation of unavoidable risks;
3 controlling hazards at source; and
4 replacing the dangerous by less or non-dangerous alternatives.

(The other principles are:

5 adapting work to the individual;
6 adapting to technical progress;
7 developing a coherent prevention policy;
8 giving priority to collective over individual protective measures; and
9 giving appropriate instructions to employees.)

(b) A series of actions arranged or organised according to rank or importance designed to control risks, which are considered in order of importance, effectiveness or priority of measures. It is designed to control risks that normally begin with an extreme measure of control and end with personal protective equipment as a last resort.

(c) The first stage of the hierarchy is to eliminate the risks by either designing them out or changing the process. The next stage is the reduction of the risks by the substitution of hazardous substances with others which are less hazardous. If this were not possible, then isolation would have to be considered, using enclosures, barriers or worker segregation. The application of engineering controls such as guarding, the provision of local exhaust ventilation systems, or the use of reduced voltage systems or residual current devices would follow, as would management controls such as safe systems of work, training, job rotation and supervision, with the final control measure being the provision of personal protective equipment such as ear defenders or respiratory protective equipment.

(d) Factors to consider when selecting risk control measures include the application of the hierarchy of control. The more serious hazards may be eliminated completely, for example by avoiding working at height with the installation of a permanent working platform with stair access. In some cases, less hazardous substances or processes may be substituted, such as the use of water-based rather than oil-based paints. The control of risks by isolating them or segregating people from the hazard is an effective control measure which is used in many instances; for example separating vehicles and pedestrians in workplaces. The rest of the hierarchy of controls may be applied in a similar way to that shown in part (c).

The evaluation of the level of risk involved, together with its likely frequency, the severity of its outcome and its extent, is another factor. Other relevant factors that should be considered include:

- the number and type of persons exposed to the risk;

- the environment in which the control is to be used;

- the requirements of the law and the recommendations contained in relevant standards and guidance;

- the existing level of competence of the persons involved, with the possible need for the provision of additional training and supervision; and

- the initial purchase and subsequent maintenance costs of the chosen risk control method.

Question 2

Following a reportable accident to a contractor while on an employer's premises, an investigation was carried out by an enforcement officer. It was found that the employer had failed to provide health and safety information to the contractor. The employer was not able to demonstrate any source of competent health and safety advice.

(a) **Outline** the duties that the employer owes to the contractor under the Health and Safety at Work etc. Act 1974. **(4)**

(b) **Outline** the health and safety information that should have been provided to the contractor before work commenced. **(8)**

(c) **Outline** the factors that the employer should consider when selecting an individual to fulfil the role of health and safety practitioner. **(8)**

Answer:

(a) The duties contained in Section 3 of the Health and Safety at Work (HSW) Act require an employer to ensure that persons working on their premises who are not direct employees, such as contractors, are not put at risk by the activities being carried out at the premises. Additionally, Section 4 places duties on the employer as a controller of premises, to ensure, for example, that the premises and its plant and equipment are safe and without risks to health for persons using them as a place of work.

(**Author comment**: It is possible to argue that the CDM Regulations are relevant here because all Health and Safety Regulations are made as a result of Section 15 of the HSW Act. However, in NEBOSH examinations, specific mention would be made of any Regulation that the candidates were expected to include in their answers. Hence this question was only concerned with the specific duties included in the Act.)

(b) The health and safety information that should be provided to the contractor before work commenced would include details of any hazards in the areas where they were to work, and details of any restricted areas and those where personal protective equipment such as hearing protection would have to be worn. Information would also be required on the emergency arrangements, such as for fire and first aid, and accident (and hazard) reporting procedures. Any general site rules, for example for signing in and out of the site, should also be supplied, and the name and location of the designated contact person. Finally, information should be provided on any training procedures for any processes or equipment to be used by the contractor.

(c) The factors to be considered when selecting an individual to fill the role of health and safety practitioner include:

■ previous training, qualifications and membership of professional bodies;

■ knowledge and understanding of the role and relevant previous experience;

■ understanding of the principles of risk assessment and relevant control methods;

■ familiarity with relevant current legislation, standards and guidance;

- personal qualities, including ability to promote and communicate good health and safety practices;

- ability to identify problems and assess the need for action – a proactive approach;

- ability to develop and implement appropriate health and safety management strategies and evaluate their effectiveness using appropriate benchmarking data; and

- awareness of personal limitations and willingness to undertake training to rectify the deficiencies.

Question 3

With reference to section 6 of the Health and Safety at Work etc. Act 1974, **outline** the general duties of designers, importers, manufacturers and suppliers of articles and substances for use at work to ensure that they are safe and without risk to health. **(8)**

Answer:

Section 6 of the Health and Safety at Work etc. Act 1974 places an obligation on persons who design, manufacture, import or supply any article or substance for use at work to ensure, so far as is reasonably practicable, that it is safe and without risk to health. Articles must be safe when they are set, cleaned, used and maintained. Substances must be without risk to health when they are used, handled, stored or transported. Testing and examination must be carried out to ensure the required level of safety, and additionally employers should be provided with information on the safe use, dismantling and disposal of the articles and substances and given revised information should a subsequent serious risk become known. Importers have a duty to ensure articles or substances comply with the requirements of UK legislation.

Question 4

(a) **Identify FOUR** common law duties that an employer must fulfil. **(4)**

(b) **Outline** practical ways in which an employer might fulfil the four duties identified in (a). **(4)**

Answer:

(a) Four common law duties that an employer must undertake are:

■ provide a safe place of work, including access and egress

■ provide safe plant and equipment

■ provide a safe system of work

■ provide safe and competent fellow employees.

An additional duty is to provide adequate levels of supervision, information, instruction and training.

(b) A safe place of work includes items such as the structure of the building and the condition of the flooring, both of which should be sound and free of defects. There should be a safe means of access and egress to the place of work, with gangways kept clear and wide enough for traffic if it is used. Environmental factors, such as working space, lighting and heating, are also important.

Safe plant and equipment will require adequate guarding or other safety devices and regular recorded maintenance.

Safe systems of work, such as permits to work and method statements, need to be available and communicated to the workforce.

Safe and competent fellow employees can be attained through pre-selection, induction and training programmes and good supervision.

As a consequence of the fifth duty, there should be adequate levels of supervision, including mentoring and monitoring of work in progress. This duty is owed to every employee as an individual, and especially to vulnerable people, including the young and those with disabilities, and remains even when the employee may be loaned to another employer.

2. NGC2 – CONTROLLING WORKPLACE HAZARDS

Question 1

Asbestos is to be removed from a building used as a workplace.

(a) **Outline** the difference between acute and chronic health effects. **(4)**

(b) **Identify** the possible chronic health risks from exposure to asbestos. **(2)**

(c) **Outline** where asbestos may be found within the building. **(4)**

(d) **Outline** control measures that should be considered before and during the removal of asbestos. **(10)**

Answer:

(a) The effects on health of hazardous substances may be either acute or chronic.

Acute effects are of short duration and appear fairly rapidly, usually during or after a single or short-term exposure to a hazardous substance. Such effects may be severe and require hospital treatment but are usually reversible. Examples include asthma-type attacks, nausea and fainting. In most cases, acute effects recede on cessation of exposure.

Chronic effects develop over a period of time which may extend over many years. The word 'chronic' means 'with time'. Chronic health effects are produced by prolonged or repeated exposures to hazardous substances, resulting in a gradual, latent and often irreversible illness, which may remain undiagnosed for many years. Many cancers and mental diseases fall into the chronic category. During the development stage of a chronic disease, the individual may experience no symptoms.

(b) Asbestos produces a fine fibrous dust of respirable dust size which can become lodged in the lungs. The fibres can be very sharp and hard, causing damage to the lining of the lungs over a period of many years. Asbestos is a possible chronic health risk. It can lead to one of the following diseases:

■ asbestosis or fibrosis (scarring) of the lungs

■ lung cancer

■ mesothelioma – cancer of the lining of the lung or, in rarer cases, the abdominal cavity.

(c) Typical sites of asbestos include ceiling tiles, corrugated asbestos cement roofing and wall sheets, sprayed asbestos coatings on structural members (to provide fire resistance), loft insulation and asbestos gaskets, asbestos cement guttering and downpipes, and the use of asbestos textiles and insulators on electric panels and fuse holders. Blue and brown asbestos are considered to be the most dangerous and may be found in older buildings, where they were used as heat insulators around boilers and hot water pipes and as fire protection of structures. White asbestos has been used in asbestos cement products and brake linings.

If asbestos is discovered during the performance of a contract, work should cease immediately and the employer should be informed.

(d) The control measures that should be considered before and during the removal of asbestos include the segregation of the area where the work is to be done and restricting access to that area. A licensed contractor must be appointed for the removal of the asbestos. A sheeted enclosure must be erected around the area to keep it under a negative pressure. There must be a restriction on the use of power tools for the removal operation, and airborne fibre levels must be monitored during the operation. All employees involved in the operation must be provided with personal protective equipment such as hooded overalls and gloves and respiratory protective equipment.

Decontamination and welfare facilities must be provided and arrangements made for cleaning the site when the removal work has been completed. At this stage, an air clearance test must be undertaken. The employees must be made aware of both the risks involved in the removal operation and the precautions to be taken.

Question 2

(a) **Identify** the possible effects on health arising from exposure to ionising radiation. **(4)**
(b) **Outline** possible means of ensuring that workers are not exposed to unacceptable levels of ionising radiation. **(4)**

Answer:

(a) Ionising radiation attacks the cells of the body by producing chemical changes in the cell DNA by ionising it (thus producing free radicals), which leads to abnormal cell growth. The effects of these ionising attacks depend on the following factors:

■ the size of the dose – the higher the dose, the more serious will be the effect;

■ the area or extent of the exposure of the body – the effects may be far less severe if only a part of the body (e.g. an arm) receives the dose;

■ the duration of the exposure – a long exposure to a low dose is likely to be more harmful than a short exposure to the same quantity of radiation.

Acute exposure can cause, dependent on the size of the dose, blood cell changes, radiation sickness (nausea and vomiting), skin burns and blistering, collapse and death. Chronic exposure can lead to anaemia, the formation of cataracts, leukaemia and other forms of cancer. It is also known that ionising radiation can have an adverse effect on the function of human reproductive organs and processes. Increases in the cases of sterility, stillbirths and malformed foetuses have also been observed.

The health effects of ionising radiation may be summarised in two categories – **somatic effects**, which refers to cell damage in the person exposed to the radiation dose, and **genetic effects**, which refers to the damage done to the children of the irradiated person.

(b) Protection is obtained by the application of shielding, time and distance, either individually or, more commonly, using a mixture of all three.

Shielding is the best method because it is an 'engineered' solution. It involves the placing of a physical shield, such as a layer of lead, steel and concrete, between the worker and the radioactive source. The thicker the shield, the more effective it is.

Time involves the use of the reduced time exposure principle and thus reduces the accumulated dose.

Distance works on the principle that the effect of radiation is reduced as the distance between the worker and the source increases. This may be achieved by remote handling and setting up restricted areas.

Other measures include the following:

- effective emergency arrangements;
- training of employees;
- the prohibition of eating, drinking and smoking adjacent to exposed areas;
- a high standard of personal cleanliness and first aid arrangements;
- strict adherence to personal protective equipment arrangements, which may include full body protection and respiratory protection equipment;
- procedures to deal with spillages and other accidents;
- prominent signs and information regarding the radiation hazards;

■ monitoring exposure by means of film badges and health surveillance of employees.

The Ionising Radiations Regulations specify a range of precautions which must be taken, including the appointment of a Radiation Protection Supervisor and a Radiation Protection Adviser.

Question 3

Identify control measures that can be implemented to reduce the risk of violence in a busy accident and emergency department. **(8)**

Answer:

A busy accident and emergency department has to balance the protection of staff from violent attack with the need to offer patients a calm environment. Protection could be given to staff by the installation of wide counters, coded locks on doors, CCTV systems, panic buttons, and personal and general alarm systems. Staff could also be issued with pagers and mobile phones. The employment of security staff and strict security procedures for the storage and issuing of drugs are two further precautions taken by such departments. Awareness training for staff so that they can recognize early signs of aggressive behaviour, and an effective counselling service for those who have suffered from violent behaviour, should be provided.

Consultation with employees or other people at risk will improve their commitment to control measures and will make the precautions much more effective. The level of awareness training and information provided, together with the general working environment, all have a significant influence on the level of risk. The training should enable staff to recognise the early signs of aggressive behaviour and avoid, defuse or otherwise deal with a violent situation.

The department should be well lit, with attention given to the décor and furnishings in an attempt to create a calm and open atmosphere. Additionally, signs should be displayed stressing the existence of a policy of zero tolerance for any form of violence. Queue-jumping causes a lot of anger and frustration and needs effective signs and proper queue management, which can help to reduce the potential for conflict. Some violence may be deterred if it is made clear to the public that any violence may be recorded and also by informing telephone callers that their calls will be recorded.

Other control measures include:

- ■ ensuring that staffing levels are adequate;

- ■ setting up support services to help victims of violence and, if necessary, other staff who could be affected. They may need debriefing, legal assistance, time off work to recover or counselling by experts; and

- ■ ensuring an effective means of communication with patients, particularly on the subject of waiting times.

NOTES

NOTES

NOTES

Test your learning with
Spot The Hazard animations

If you would like extra help preparing for your assessment then take a look at these handy revisions tools!

www.routledge.com/cw/hughes

Available on our companion website these online animations will help you identify common instances of bad practice.

See if you can spot the hazards and then find out if you're right. All instances of bad practice are fully explained, the correct practice is then shown with reference to legislation and theory.

Our free companion website supports our range of NEBOSH books and also provides access to:

- Editable and printable health and safety forms

- Fully downloadable image library taken from the companion textbooks

- Videos of the authors

www.routledge.com/cw/hughes